Welcome to FOSS Next Generation Middle School Program Features

The Next Generation of *Active Learning*™

FOSS Next Generation Middle School represents more than 20 years of working with teachers and students with the goal of finding the most effective ways to help all teachers teach science and all students learn science.

FOSS and the Next Generation Science Standards (NGSS)

FOSS Next Generation Middle School engages students at the intersection of the disciplinary core ideas, science and engineering practices, and crosscutting concepts. FOSS knows how to integrate the NGSS into your classroom.

Put our experience to work for you.

Features for the Next Generation Middle School Program

- Alignment to NGSS
- Strategies for developing disciplinary core ideas, engaging in science and engineering practices, and exposing crosscutting concepts
- Strategies for supporting the CCSS Literacy in Science and Technology for Middle School
- Solving real-world problems to engage students in real-world applications, engineering opportunities, research, dealing with local and regional environmental issues, and more
- Technology resources including tablet-based Investigation Guides
- Six and twelve week courses
- Professional development support

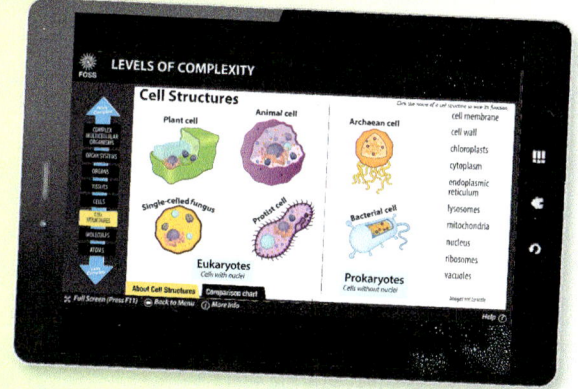

Explore FOSS Next Generation Middle School and see how it embodies what the NGSS describes:

Understanding of science develops over time—
FOSS has elaborated learning progressions that result in more sophisticated ways of thinking about core ideas over multiple years.

Focus on in-depth explorations of fewer topics—
FOSS modules provide deep engagement with the important science ideas, resulting in more effective learning outcomes than when many topics are only briefly visited.

Science is more than a body of knowledge—
FOSS embraces the belief that science content cannot be separated from its practices. Every FOSS module provides opportunities for students to engage in and understand science and engineering practices while exploring the disciplinary core ideas.

FOSS Next Generation Middle School Program Features

The FOSS Program Design

FOSS includes a wide range of experiences that help build student understanding of science concepts and scientific habits of mind. Students are actively engaged and have ownership of their learning in everything they do with FOSS Next Generation Middle School.

The Elements of Active Middle School Investigation

Active Investigation

- Using Formative Assessment
- Integrating Science Notebooks
- Solving Real-World Problems and Engineering Challenges
- Engaging in Science–Centered Language Development
- Engaging with Technology
- Reading FOSS Science Resources Books

Full Option Science System

FOSStering Active Investigation

Our research-based FOSS Investigations Guide helps you teach using effective pedagogical practices in every science lesson.

Here's how it works:

1. **Plan Thoughtfully with Confidence**
   The Investigations Guide is your core resource for facilitating the active investigations in each course. You can teach with confidence, knowing the activities reflect extensive field-testing experiences with teachers in diverse classrooms. The Investigations Guide is available in hard copy print and digital format online.

2. **Facilitate the Science Investigation**
   Whether you're a novice or an experienced veteran, FOSS guides you efficiently through the investigations. Using the detailed support, guiding questions, and differentiation strategies, you'll be in control every step of the way.

3. **Incorporate Learning in Engineering Design**
   In many FOSS Middle School courses, students learn and exercise the concepts and skills of engineering design. In various contexts, FOSS investigations provide opportunities for students to learn to define a problem, develop and test possible solutions, analyze data from tests, and revise the best ideas into a satisfactory solution.

4. **Provide Relevant Experiences for Students**
   FOSS is designed to help students become effective at planning and carrying out investigations to answer questions, and to discover solutions to problems that build on their K–5 experiences. The investigations are organized as a progression in which students learn to test and develop supporting evidence for scientific models.

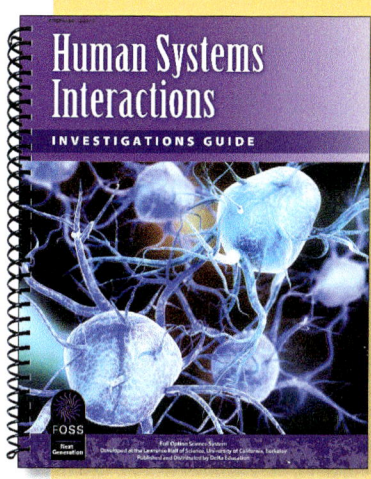

FOCUS QUESTION

How do cells in the human body get the resources they need?

Each part of each investigation is driven by a focus question, which identifies the challenge to be met or problem to be solved.

Recommended FOSS Next Generation Scope and Sequence

Grade	Physical Science	Earth Science	Life Science
6–8	Waves* Gravity and Kinetic Energy*	Planetary Science	Heredity and Adaptation* Human Systems Interactions*
	Chemical Interactions	Earth History	Populations and Ecosystems
	Electromagnetic Force* Variables and Design*	Weather and Water	Diversity of Life
5	Mixtures and Solutions	Earth and Sun	Living Systems
4	Energy	Soils, Rocks, and Landforms	Environments
3	Motion and Matter	Water and Climate	Structures of Life
2	Solids and Liquids	Pebbles, Sand, and Silt	Insects and Plants
1	Sound and Light	Air and Weather	Plants and Animals
K	Materials and Motion	Trees and Weather	Animals Two by Two

*Half-length course

FOSS Next Generation Middle School Program Features

Connecting Science and Language Arts

In FOSS, students experience the natural world in authentic ways and use language to inquire, process information, and communicate their thinking. As students read, write, listen, and speak about the concepts they explore, and engage in argument from evidence, they build Common Core State Standards (CCSS) literacy skills. FOSS Next Generation utilizes multiple strategies and resources to reinforce the CCSS for English Language Arts.

Informational Text

Reading is an integral part of science—both in the classroom and for students' potential careers. Common Core calls for teachers to infuse the science curriculum with opportunities for students to engage with discipline-specific nonfiction texts as a means for acquiring additional information and new ideas.

FOSS Science Resources

FOSS Science Resources books provide students with important details connected to the hands-on investigations in a popular science article format. Readers use the information to draw logical inferences and conclusions, or extend the themes of the text to other contexts.

The text and graphics help readers analyze important course concepts in order to draw conclusions about the texts' meaning, and also to compare and contrast to other sources of information.

FOSS Science Resources books are available in two formats; digital and print.

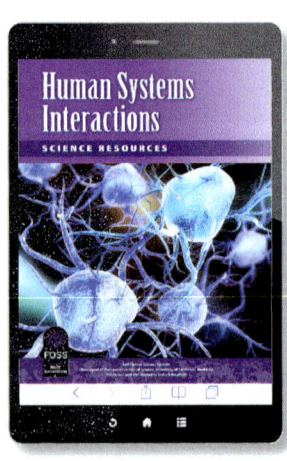

eBooks are accessible on mobile devices.

Unleash the power of student writing and discourse

When students engage in informal writing, their ability to process information is enhanced. Writing informational or explanatory text promotes the use of science and engineering practices and develops a deeper engagement with the science content.

Science Notebooks

Keeping a notebook helps students organize observations, process data, and maintain a record of their learning for future reference. The processes of writing about science experiences and communicating thinking are powerful learning devices for students.

- Notebook entries form one of the core elements of the assessment system
- Students craft explanatory narratives to make sense of their science experiences
- Science notebook prompts and masters are provided for facilitation of student writing and discussion

FOSS Next Generation Middle School Program Features

Tap into Technology

FOSSweb is the gateway to online learning and teaching resources.
Here is a sampling of the many resources you'll find:

eInvestigations Guides provide teachers access wherever you go.

Videos provide detailed investigation information.

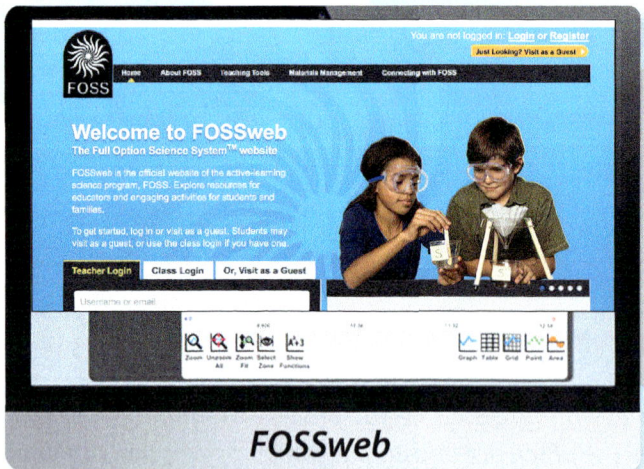

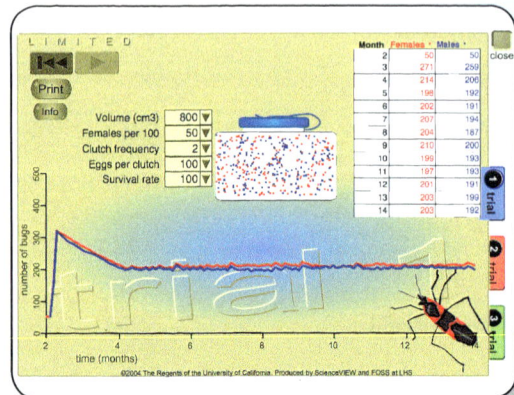

Online simulations provide opportunities for students to continue experimenting beyond the scope of the classroom walls.

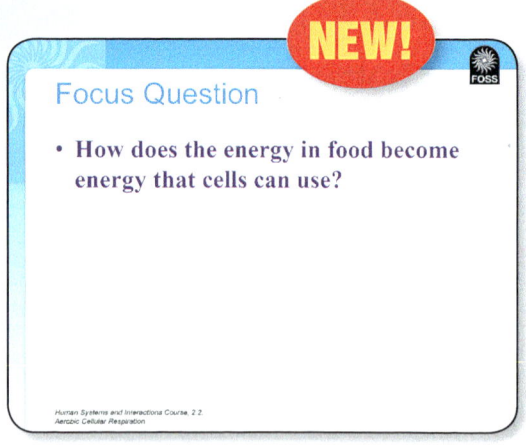

Teaching Slides help teachers and students during the investigations.

Reliable Assessment

FOSS provides teachers with a research-based assessment system.

The **FOSS** Assessment system for Middle School is designed to be used to monitor progress during the investigation and as evaluation tools at the end of the course.

The **FOSS** assessment system features:

- Embedded assessments provide formative and diagnostic information during each part of an investigation.
- Strategies for reviewing notebook entries and for observing students engaged in science practices.
- Performance assessments that support teachers in assessing student learning across the three dimensions of content, practices, and crosscutting concepts.

*The **FOSS** assessment system was developed with partial support from the National Science Foundation.*

FOSSmap Online Assessment and Reporting

FOSSmap is the new online assessment management solution for FOSS.

- Students can take **FOSS** assessments online
- Automatic coding provides timely feedback to the teacher
- Monitor class progress and inform instruction
- Reports help you identify which students need more help with specific concepts

FOSS provides the right tools for the job!

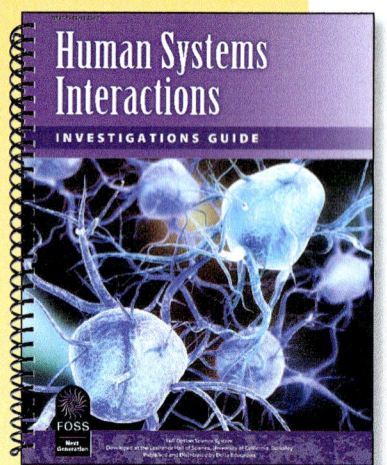

Investigations Guide
The research-based strategies in this core instructional tool provide you with unprecedented support that will allow you to easily manage learning around both the concepts and practices of science and engineering.

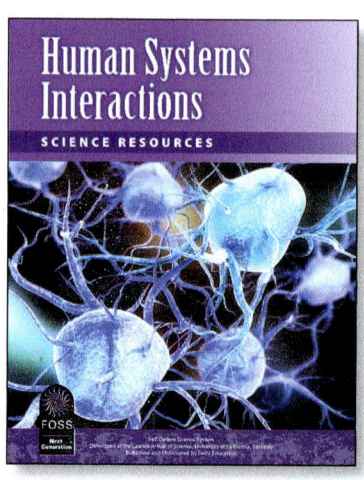

FOSS Science Resource Books
These student readings connect students' first-hand experiences to informational text.

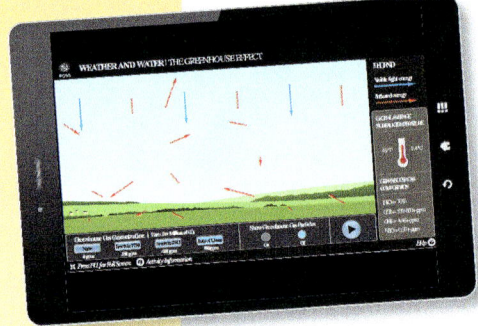

FOSSweb
The redesigned FOSSweb provides you with even more resources to engage students. Our tablet-ready eInvestigations Guides put the support you need right at your fingertips in an easy-to-navigate digital format. There are also streaming instructional videos, teaching slides, FOSSmap assessment, teacher preparation videos, and much more.

Teacher Resources
This tool guides teachers in how the elements of FOSS work, including language strategies for science teachers, and provides notebook and assessment masters.

Equipment Kit
Specifically-designed materials in FOSS kits lead to successful investigations for all students. Our new kits are formatted with more options to meet your specific needs.

PUBLISHED AND DISTRIBUTED BY

DEVELOPED AT

INVESTIGATIONS GUIDE

Human Systems Interactions

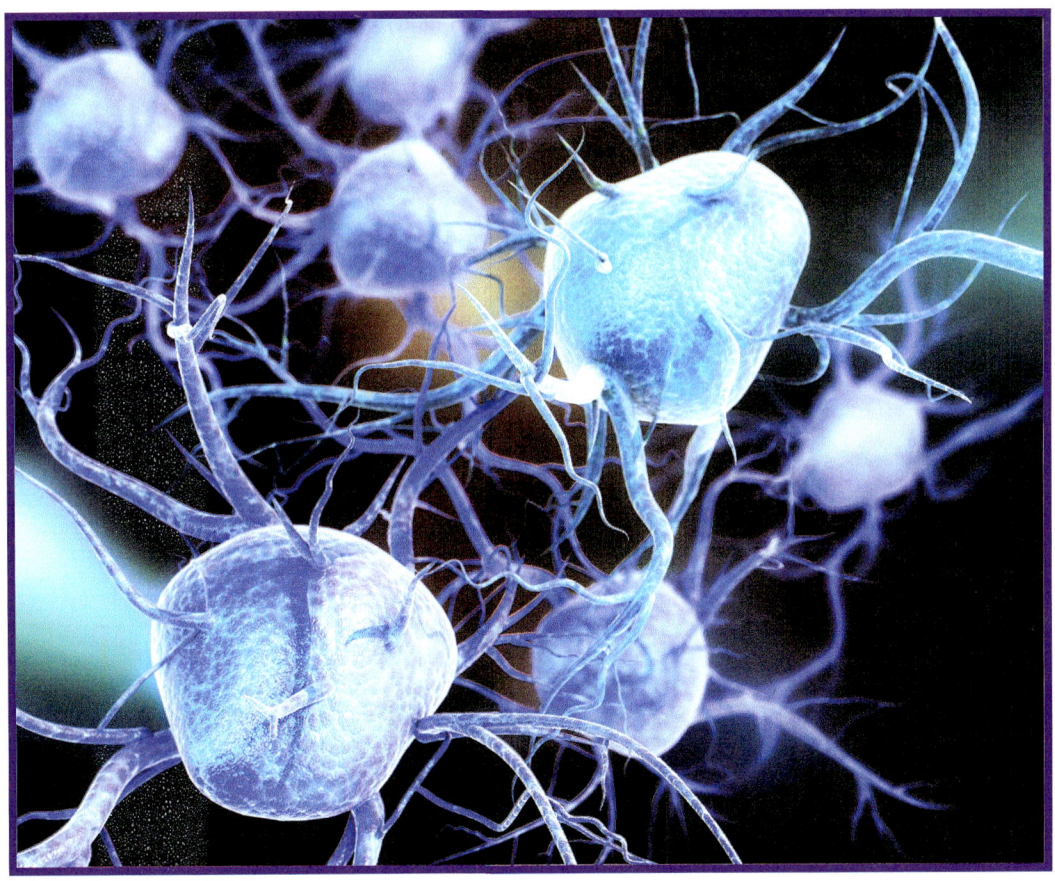

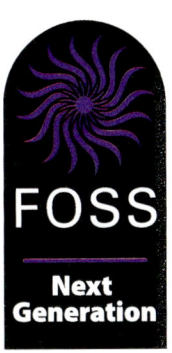

Full Option Science System
Developed at the Lawrence Hall of Science, University of California, Berkeley
Published and Distributed by Delta Education

FOSS Lawrence Hall of Science Team
Larry Malone and Linda De Lucchi, FOSS Project Codirectors and Lead Developers Jessica Penchos, Middle School Coordinator; Kathy Long, FOSS Assessment Director; David Lippman, Program Manager; Carol Sevilla, Publications Design Coordinator; Susan Stanley, Graphic Production; Rose Craig, Illustrator
FOSS Curriculum Developers: Alan Gould, Teri Lawson, Ann Moriarty, Virginia Reid, Terry Shaw, Joanna Snyder
FOSS Technology Developers: Susan Ketchner, Arzu Orgad
FOSS Multimedia Team: Kate Jordan, Senior Multimedia; Christopher Keller, Multimedia Producer; Jonathan Segal, Designer; Christopher Cianciarulo, Designer; Dan Bluestein, Programmer; Shan Jiang, Programmer

Delta Education Team
Bonnie A. Piotrowski, Editorial Director, Elementary Science
Project Team: Jennifer Apt, Sandra Burke, Mary Connell, Joann Hoy, Angela Miccinello, Jennifer Staltare

Content Reviewer
Rocco Carsia, Associate Professor
Department of Cell Biology
School of Osteopathic Medicine
Rowan University, New Jersey

Thank you to all FOSS Middle School Revision Trial Teachers and District Coordinators
Frances Amojioyi, Lincoln Middle School, Alameda, CA; Dean Anderson, Organized trials for Boston Public Schools, Boston, MA; Thomas Archer, Organized trials for ESD 112, Vancouver, WA; Lauresa Baker, Lincoln Middle School, Alameda, CA; Bobbi Anne Barnowsky, Canyon Middle School, Castro Valley, CA; Christine Bertko, St. Finn Barr Catholic School, San Francisco, CA; Stephanie Billinge, James P. Timilty Middle School, Roxbury, MA; Jerry Breton, Ingleside Middle School, Phoenix, AZ; Robert Cho, Timilty Middle School, Boston, MA; Susan Cohen, Cherokee Heights Middle School, Madison, WI; Malcolm Davis, Canyon Middle School, Castro Valley, CA; Marilyn Decker, Organized trials for Milton PS, Milton, MA; Jenny Ernst, Park Day School, Oakland, CA; Marianne Floyd, Spanaway Middle School, Spanaway, WA; Sarah Kathryn Gessford, Journeys School, Jackson, WY; Charles Hardin, Prairie Point Middle School, Cedar Rapids, IA; Jennifer Hartigan, Lincoln Middle School, Alameda, CA; Sheila Holland, TechBoston Academy, Boston, MA; Nicole Hoyceanyls, Charles S. Pierce Middle School, Milton, MA; Bruce Kamerer, Donald McKay K-8 School, East Boston, MA; Carmen Saele Kardokus, Reeves Middle School, Olympia, WA; Janey Kaufman, Organized trials for Scottsdale USD, Scottsdale, AZ; Erica Larson, Organized trials for Grant Wood AEA, Cedar Rapids, IA; Lindsay Lodholz, O'Keeffe Middle School, Madison, WI; Robert Mattisinko, Chaparral High School, Scottsdale, AZ; Brenda McGurk, Prairie Point Middle School, Cedar Rapids, IA; Tim Miller, Mountainside Middle School, Scottsdale, AZ; Thomas Miro, Lincoln Middle School, Alameda, CA; Spencer Nedved, Frontier Middle School, Vancouver, WA; Joslyn Olsen, Lincoln Middle School, Alameda, CA; Stephanie Ovechka, Cedarcrest Middle School, Spanaway, WA; Barbara Reinert, Copper Ridge School, Scottsdale, AZ; Stephen Ramos, Lincoln Middle School, Alameda, CA; Gina Rutenbeck, Prairie Point Middle School, Cedar Rapids, IA; John Sheridan, Boston Public Schools (Boston Schoolyard Initiative), Boston, MA; Barbara Simon, Timilty Middle School, Boston, MA; Lise Simpson, Alcott Middle School, Norman, OK; Autumn Stevick, Thurgood Marshall Middle School, Olympia, WA; Ted Stoeckley, Hall Middle School, Larkspur, CA; Lesli Taschwer, Organized trials for Madison SD, Madison, WI; Paula Warner, Alcott Middle School, Norman, OK; Darren T. Wells, James P. Timilty Middle School, Boston, MA; Kristin White, Frontier Middle School, Vancouver, WA

Photo Credits: © MichaelTaylor3d/Shutterstock (cover); © Lightspring/Shutterstock; © Laurie Meyer/Delta Education; © UGREEN 3S/Shutterstock; © iStockphoto/Eraxion; © ktsdesign/Shutterstock; © Delta Education

Published and Distributed by Delta Education, a member of the School Specialty Family
The FOSS program was developed in part with the support of the National Science Foundation grant nos. ESI-9553600 and ESI-0242510. However, any opinions, findings, conclusions, statements, and recommendations expressed herein are those of the authors and do not necessarily reflect the views of NSF. FOSSmap was developed in collaboration between the BEAR Center at UC Berkeley and FOSS at the Lawrence Hall of Science. Score analysis is done through the BEAR Center Scoring Engine.

Copyright © 2017 by The Regents of the University of California

All rights reserved. Any part of this work (other than duplication masters) may not be reproduced or transmitted in any form or by any means, electronic or mechanical, including photocopying and recording, or by an information storage or retrieval system without permission of the University of California. For permission please write to: FOSS Project, Lawrence Hall of Science, University of California, Berkeley, CA 94720.

Human Systems Interactions — Teacher Toolkit, 1465646
Investigations Guide, 1465686
978-1-62571-199-1
Printing 1 – 3/2016
Webcrafters, Madison, WI

INVESTIGATIONS GUIDE

Human Systems Interactions

TABLE OF CONTENTS

Overview	1
Framework and NGSS	29
Materials	49
Investigation 1: Systems Connections	59
Part 1: Human Body Structural Levels	72
Part 2: Systems Research	83
Investigation 2: Supporting Cells	99
Part 1: Food and Oxygen	110
Part 2: Aerobic Cellular Respiration	123
Investigation 3: The Nervous System	143
Part 1: The Sense of Touch	156
Part 2: Sending a Message	175
Part 3: Other Senses	199
Part 4: Learning and Memory	216
Assessment	237

FOSS FORWARD THINKING

The FOSS Vision

When the Full Option Science System (FOSS) began, the founders envisioned a science curriculum that was enjoyable, logical, and intuitive for teachers, and stimulating, provocative, and informative for students. Achieving this vision was informed by research in cognitive science, learning theory, and critical study of effective practice. The modular design of the FOSS product allowed users to select topics that aligned with district or state learning objectives, or simply resonated with their perception of comprehensive and reasonable science instruction. The original design of the FOSS Program was comprehensive in terms of coverage. FOSS was designed to provide real and meaningful student experience with important scientific ideas and to nurture developmentally appropriate knowledge of the objects, organisms, systems, and principles governing, the natural world.

The FOSS Next Generation Revision

But the developers never envisioned FOSS to be a static curriculum, and now the Full Option Science System has evolved into a fully realized 21st century science program with authentic connection to the *Next Generation Science Standards (NGSS)*. The FOSS science curriculum is a comprehensive science program, featuring instructional guidance, student equipment, student reading materials, digital resources, and an embedded assessment system. FOSS has always utilized an inquiry approach to teaching and learning, but the *Framework for K–12 Science Education*, on which the NGSS are based, has provided a new way for the FOSS developers to think about and communicate the FOSS message. The FOSS philosophy has always taken very seriously the teaching of good, comprehensive, accurate, science content using the methods of inquiry to advance that science knowledge. But the *Framework* has allowed us to articulate our mission in a more coherent manner, using the vocabulary established by the authors of the *Framework*. The FOSS instructional design now strives to

a. communicate the disciplinary core ideas (content) of science, while

b. guiding and encouraging students to engage in or exercise the science and engineering practices (inquiry methods) to develop knowledge of the disciplinary core ideas, and

c. help students apprehend the crosscutting concepts (themes that unite core ideas, overarching concepts) that connect the learning experiences within a discipline and bridge meaningfully across disciplines as students gain more and more knowledge of the natural world.

> "Full Option Science System has evolved into a fully realized 21st century science program with authentic connection to the Next Generation Science Standards (NGSS). The FOSS science curriculum is a comprehensive science program, featuring instructional guidance, student equipment, student reading materials, digital resources, and an embedded assessment system."

Full Option Science System

FOSS FORWARD THINKING

> *"FOSS is crafted with a structured, yet flexible, teaching philosophy that embraces the much-heralded 21st century skills; collaborative teamwork, critical thinking, and problem solving."*

The NGSS describe the knowledge and skills we expect our students to be able to demonstrate after completing their science instruction experience. The expectations are demanding and include no small measure of ability to communicate scientific knowledge. The ability to communicate complex ideas assumes that students have had a significant amount of experience and practice building coherent explanations, defending claims, and organizing and presenting reasoned arguments in the context of their science curriculum. This is where scientific inquiry encounters language arts. FOSS draws on both the Common Core State Standards (CCSS) for English Language Arts and research data regarding the productive use of student science notebooks. FOSS developers realize that the most effective science program must seamlessly integrate science instruction goals and language arts skills. Science is one of the most engaging and productive arenas for introducing and exercising language arts skills: vocabulary, nonfiction (informational) reading, cause-and-effect relationships, on and on.

FOSS is strongly grounded in the realities of the classroom and the interests and experiences of the learners. The content in FOSS is teachable and learnable over multiple grade levels as students increase in their abilities to reason about and integrate complex ideas within and between disciplines.

FOSS is crafted with a structured, yet flexible, teaching philosophy that embraces the much-heralded 21st century skills; collaborative teamwork, critical thinking, and problem solving. The FOSS curriculum design promotes a classroom culture that allows both teachers and students to assume prominent roles in the management of the learning experience.

FOSS is built on the assumptions that understanding of core scientific knowledge and how science functions is essential for citizenship, that all teachers can teach science, and that all students can learn science. Formative assessment in FOSS creates a community of reflective practice. Teachers and students make up the community and establish norms of mutual support, trust, respect, and collaboration. The goal of the community is that everyone will demonstrate progress and will learn and grow.

Full Option Science System

HUMAN SYSTEMS INTERACTIONS — *Overview*

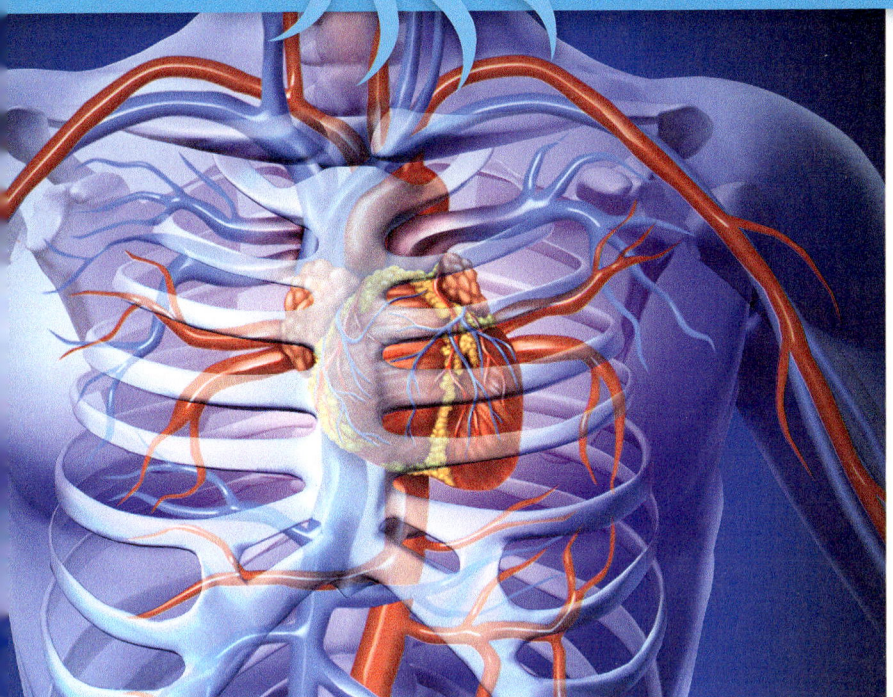

INTRODUCTION

Life is a complex of interactions; human life is no exception. The basis of the human body is the cell. Associations of cells work together to form tissues, which form organs. Organs work together to perform specific functions in organ systems. And finally, the array of organ systems make up a human body.

Middle school students are prepared to explore how organ systems interact to support each and every cell in the body. What happens when the body is attacked by an invader or an organ system malfunctions? How do cells get the resources they need to live? How do cells gain access to the energy stored in energy-rich compounds? How do systems support the human organism as it senses and interacts with the environment?

These questions inspire students to find out more, and may spawn a lifetime of learning about their body systems and the environmental factors that affect them. Questions like these have the potential to help students understand and appreciate what may be of highest importance to them, themselves.

FOSS Human Systems Interactions is a 5-week course that can be bundled with the **FOSS Diversity of Life Course** or used as a stand-alone course.

Contents

Introduction	1
Course Matrix	2
FOSS Middle School Components	4
FOSS Instructional Design	8
Differentiated Instruction	16
FOSS Investigation Organization	18
Management Strategies	20
Working in Collaborative Groups	25
Safety in the Classroom	27
FOSS Contacts	28

The NGSS Performance Expectations addressed in this module include:

Life Sciences
MS-LS1-1
MS-LS1-3
MS-LS1-7
MS-LS1-8

Full Option Science System

HUMAN SYSTEMS INTERACTIONS — Overview

	Investigation Summary	Time	Parts and Focus Questions
Inv. 1	**Systems Connections** Students solve a disease mystery. On the path to diagnosis, students discover the structural levels in human bodies: that cells form tissues, tissues form organs, organs form organ systems, and systems form a complex multicellular organism, the human. They look for evidence of how the organ systems interact, each dependent on all the others for its needs.	**Activities** 5 sessions * **Assessment** 1 session	Part 1 **Human Body Structural Levels**, 1 session *What is a human body made of?* Part 2 **Systems Research**, 4 sessions *How do human organ systems interact?*
Inv. 2	**Supporting Cells** Students fatigue their muscles and think about how their cells obtain the food and oxygen they need from the digestive, respiratory, and circulatory systems. They learn how aerobic cellular respiration works in cells. They find out that the cells eliminate wastes produced during aerobic cellular respiration via circulatory, respiratory, and excretory systems.	**Activities** 5 sessions **Assessment** 2 sessions	Part 1 **Food and Oxygen**, 3 sessions *How do cells in the human body get the resources they need?* Part 2 **Aerobic Cellular Respiration**, 2 sessions *How does the energy in food become energy that cells can use?*
Inv. 3	**The Nervous System** Students explore the different senses to understand how humans acquire information from the environment. They engage in a "neuron relay" to model how sensory information travels to the brain for processing and how information returns to the body for action. Students turn their attention to their own learning and memory formation.	**Activities** 11 sessions **Assessment** 3–4 sessions	Part 1 **The Sense of Touch**, 2 sessions *How doed the sense of touch work in humans?* Part 2 **Sending a Message**, 3 sessions *How do messages travel to and from the brain?* Part 3 **Other Senses**, 3 sessions *How are senses alike and how are they different?* Part 4 **Learning and Memory**, 3 sessions *How do humans learn and form memories?*

* A class session is 45–50 minutes

Course Matrix

Content and Disciplinary Core Ideas	Literacy/Technology	Assessment
• Multicellular organisms are complex systems composed of organ systems, which are made of organs, which are made of tissues, which are made of cells. • Cells are made of cell structures, which are made of molecules, which are made of atoms. • The human body is a system of interacting subsystems (circulatory, digestive, endocrine, excretory, muscular, nervous, respiratory, skeletal).	**Science Resources Book** "Human Organ Systems" "Disease Information" **Online Activities** "Levels of Complexity" (optional) "Human Body Structural Levels" **Video** Doctor Interviews 1 and 2	**Benchmark Assessment** *Entry-Level Survey* **NGSS Performance Expectations** MS-LS1-1 (foundational) MS-LS1-3
• The human body is a system of interacting subsystems. • The respiratory system supplies oxygen and the digestive system supplies energy (food) to all the cells in the body. • The circulatory system transports food and oxygen to the cells in the body and carries waste products to the excretory/respiratory systems for disposal. • Aerobic cellular respiration is the process by which energy stored in food molecules is converted into usable energy for cells.	**Science Resources Book** Aerobic Cellular Respiration" **Online Activity** "Human Cardiovascular System" **Videos** *Digestive and Excretory Systems* *Circulatory and Respiratory Systems*	**Benchmark Assessment** *Investigations 1–2 I-Check* **NGSS Performance Expectations** MS-LS1-3 MS-LS1-7
• Sensory receptors respond to any array of mechanical, chemical, and electromagnetic stimuli. • Sensory information is transmitted electrically to the brain along neural pathways for processing and response. • Neural pathways change and grow as information is acquired and stored as memories.	**Science Resources Book** "Sensory Receptors" "Touch" "Hearing" "Sensory Activity Brain Map" "Brain Messages" "Neurotransmission" "Smell and Taste" "Sight" "Memory and Your Brain" **Online Activities** "Touch Menu" "Brain: Synapse Function" "Brain: Neuron Growth" "Smell Menu" "Vision Menu" "Reaction Timer" **Video** *How Memory Works*	**Benchmark Assessment** *Investigation 3 I-Check* *Posttest* **NGSS Performance Expectations** MS-LS1-3 MS-LS1-8

Human Systems Interactions Course—FOSS Next Generation

HUMAN SYSTEMS INTERACTIONS — Overview

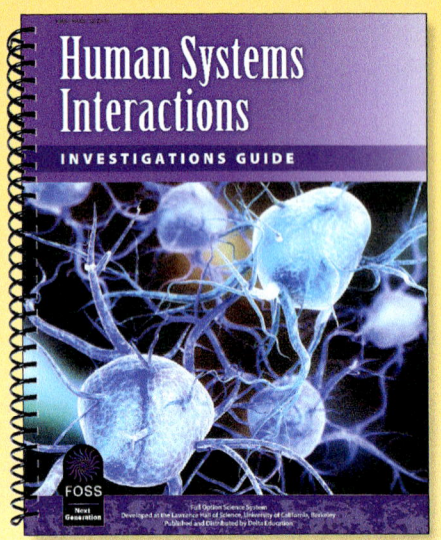

FOSS MIDDLE SCHOOL COMPONENTS

Teacher Toolkit

Each course comes with a *Teacher Toolkit*. The *Teacher Toolkit* is the most important part of the FOSS Program. It is here that all the wisdom and experience contributed by hundreds of educators have been assembled. Everything we know about the content of the course, how to teach the subject, and the resources that will assist the effort are presented here. Each middle school toolkit has three parts.

Investigations Guide. This spiral-bound document contains these chapters.

- Overview
- Framework and NGSS
- Materials
- Investigations (three in this course)
- Assessment

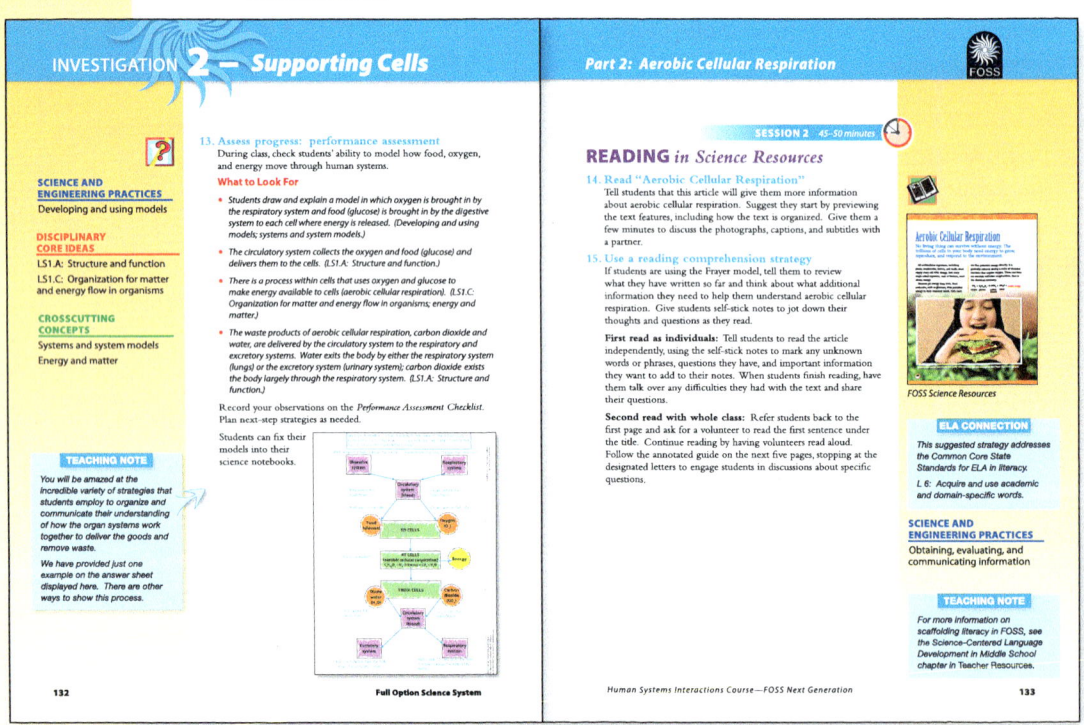

FOSS Science Resources book. One copy of the student book of readings is included in the *Teacher Toolkit*.

4 Full Option Science System

FOSS Components

Teacher Resources. These chapters can be downloaded from FOSSweb and are in the bound *Teacher Resources* book.

- FOSS Program Goals
- Science Notebooks in Middle School
- Science-Centered Language Development in Middle School
- FOSSweb and Technology
- Science Notebook Masters
- Teacher Masters
- Assessment Masters
- Notebook Answers

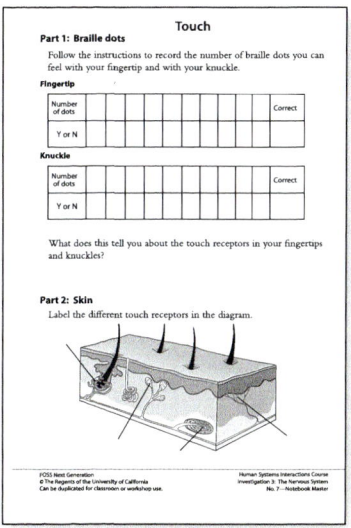

Equipment for Each Course

The FOSS Program provides the materials needed for the investigations in sturdy, front-opening drawer-and-sleeve cabinets. Inside, you will find high-quality materials packaged for a class of 32 students. Consumable materials are supplied for five sequential uses (five periods in one day) before you need to resupply. You will need to supply some items usually available in middle school science classrooms and they are listed separately in the materials lists.

The middle school equipment kits are divided into unique permanent items, common permanent items, and consumable items. Speak to your FOSS sales representative about custom configuration to best address your classroom needs.

Human Systems Interactions Course—FOSS Next Generation

HUMAN SYSTEMS INTERACTIONS — Overview

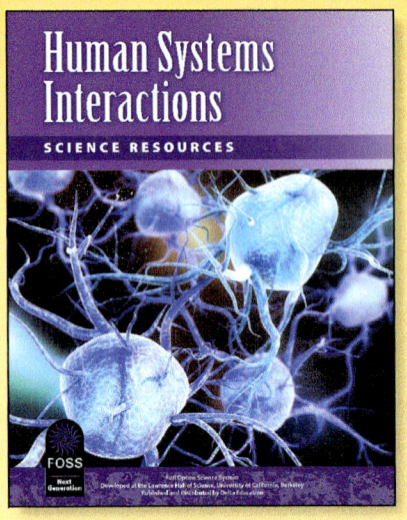

FOSS Science Resources Books

FOSS Science Resources: Human Systems Interactions is a book of original readings developed to accompany this course. The readings are referred to as articles in the *Investigations Guide*. Students read the articles in the book as they progress through the course. Some readings are done in class, others are used as homework or for research projects. The articles cover specific concepts, usually after the concepts have been introduced in the active investigation.

The articles in *FOSS Science Resources* and the discussion questions provided in *Investigations Guide* help students make connections to the science concepts introduced and explored during the active investigations. Concept development is most effective when students are allowed to experience organisms, objects, and phenomena firsthand before engaging the concepts in text. The text and illustrations help make connections between what students experience concretely and the ideas that explain their observations.

Human Organ Systems

The sections in this article introduce one of the truly exceptional systems in the world. That system is your body.

Your body is essentially the same as all of the other humans in the world. Every body is a system composed of the same fundamental organ subsystems. Most of them will be somewhat familiar to you. They are the **circulatory system**, **digestive system**, **endocrine system**, **excretory system**, **muscular system**, **nervous system**, **respiratory system**, and **skeletal system**.

You are about to begin an exploratory experience into the structure and function of human organ systems. You will become a specialist in one of these systems. You will develop knowledge about how one system functions and how it interacts with each other system. And you will become aware of **symptoms** that signal a problem with the system. This information will help you work with other system specialists to diagnose a problem.

Complex, interacting organ systems work together to make every human action and activity possible, from reading this page to taking a heart-pounding run on the beach.

Investigation 1: Systems Connections

Full Option Science System

FOSS Components

Technology

The FOSS website opens new horizons for educators and students in the classroom or at home. Each module has digital resources for students—interactive simulations, resources for research, and online activities. For teachers, FOSSweb provides resources for materials management, general teaching tools for FOSS, purchasing links, contact information for the FOSS Program, and technical support. You do not need an account to view this general FOSS Program information. In addition to the general information, FOSSweb provides PDF versions of the *Teacher Resources* component of the *Teacher Toolkit* and digital-only resources that supplement the print and kit materials.

Additional resources are available to support FOSS teachers. With an educator account, you can customize your homepage, set up easy access to the digital components of the courses you teach, and create class pages for your students with access to multimedia and online assessments.

▶ **NOTE**
To access all the teacher resources and to set up customized pages for using FOSS, log in to FOSSweb through an educator account. See the FOSSweb and Technology chapter in *Teacher Resources* for more specifics.

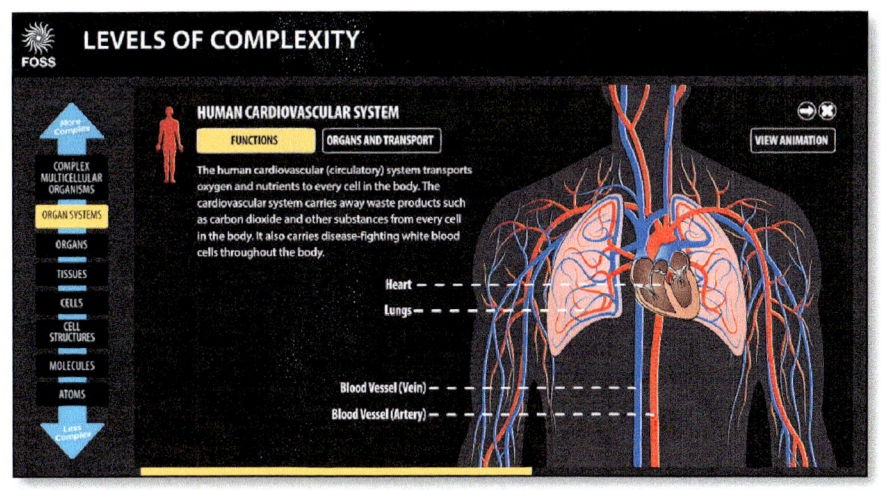

Ongoing Professional Learning

The Lawrence Hall of Science and Delta Education strive to develop long-term partnerships with districts and teachers through thoughtful planning, effective implementation, and ongoing teacher support. FOSS has a strong network of consultants who have rich and experienced backgrounds in diverse educational settings using FOSS.

▶ **NOTE**
Look for professional-development opportunities and online teaching resources on www.fossweb.com.

Human Systems Interactions Course—FOSS Next Generation

HUMAN SYSTEMS INTERACTIONS — Overview

FOSS INSTRUCTIONAL DESIGN

FOSS is designed around active investigations that provide engagement with science concepts and science and engineering practices. Surrounding and supporting those firsthand investigations are a wide range of experiences that help build student understanding of core science concepts and deepen scientific habits of mind.

- Using Formative Assessment
- Integrating Science Notebooks
- Active Investigation
- Solving Real-World Problems and Engineering Challenges
- Engaging in Science–Centered Language Development
- Engaging with Technology
- Reading FOSS Science Resources Books

FOSS Instructional Design

Each FOSS investigation follows a similar design to provide multiple exposures to science concepts. The design includes these pedagogies.

- Active investigation in collaborative groups: firsthand experiences with objects, organisms, and materials in the natural and designed worlds
- Recording in science notebooks to answer the focus question
- Reading in *FOSS Science Resources* books
- Online activities to gather data or information or to extend the investigation
- Opportunities to apply knowledge to solve problems through the engineering design process or to address regional ecological issues
- Assessment to monitor progress and motivate student learning

In practice, these components are seamlessly integrated into a curriculum designed to maximize every student's opportunity to learn. An instructional sequence may move from one pedagogy to another and back again to ensure adequate coverage of a concept.

A **learning cycle** is an instructional model based on a constructivist perspective that calls on students to be actively involved in their own learning. The model systematically describes both teacher and learner behaviors in a systematic approach to science instruction.

The most recent model is a series of five phases of intellectual involvement known as the 5Es: engage, explore, explain, elaborate, and evaluate. The body of foundational knowledge that informs contemporary learning-cycle thinking has been incorporated seamlessly and invisibly into the FOSS curriculum design.

HUMAN SYSTEMS INTERACTIONS — Overview

Active Investigation

Active investigation is a master pedagogy. Embedded within active learning are a number of pedagogical elements and practices that keep active investigation vigorous and productive. The enterprise of active investigation includes

- context: questioning and planning;
- activity: doing and observing;
- data management: recording, organizing, and processing;
- analysis: discussing and writing explanations.

Context: questioning and planning. Active investigation requires focus. The context of an inquiry can be established with a focus question or challenge from you, or in some cases, from students—How do human organs interact? At other times, students are asked to plan a method for investigation. This might include determining the important data to gather and the necessary tools. In either case, the field available for thought and interaction is limited. This clarification of context and purpose results in a more productive investigation.

Activity: doing and observing. In the practice of science, scientists put things together and take things apart, they observe systems and interactions, and they conduct experiments. This is the core of science—active, firsthand experience with objects, organisms, materials, and systems in the natural world. In FOSS, students engage in the same processes. Students often conduct investigations in collaborative groups of four, with each student taking a role to contribute to the effort.

The active investigations in FOSS are cohesive, and build on each other and the readings to lead students to a comprehensive understanding of concepts. Through the investigations, students gather meaningful data.

Online activities throughout the course provide students with opportunities to collect data, manipulate variables, and explore models and simulations beyond what can be done in the classroom. Seamless integration of the online activities forms an integral part of students' active investigations in FOSS.

Data management: recording, organizing, and processing. Data accrue from observation, both direct (through the senses) and indirect (mediated by instrumentation). Data are the raw material from which scientific knowledge and meaning are synthesized. During and after work with materials, students record data in their notebooks. Data recording is the first of several kinds of student writing.

FOSS Instructional Design

Students then organize data so that they will be easier to think about. Tables allow efficient comparison. Organizing data in a sequence (time) or series (size) can reveal patterns. Students process some data into graphs, providing visual display of numerical data. They also organize data and process them in the science notebook.

Analysis: discussing and writing explanations. The most important part of an active investigation is extracting its meaning. This constructive process involves logic, discourse, and existing knowledge. Students share their explanations for phenomena, using evidence generated during the investigation to support their ideas. They conclude the active investigation by writing in their notebooks a summary of their learning as well as questions raised during the activity.

Science Notebooks

Research and best practice have led us to place more emphasis on the student science notebook. Keeping a notebook helps students organize their observations and data, process their data, and maintain a record of their learning for future reference. The process of writing about their science experiences and communicating their thinking is a powerful learning device for students. And the student notebook entries stand as a credible and useful expression of learning. The artifacts in the notebooks form one of the core elements of the assessment system.

You will find the duplication masters for middle school presented in a notebook format. They are reduced in size (two copies to a standard sheet) for placement (glue or tape) in a bound composition book. Student work is entered partly in spaces provided on the notebook sheets and partly on adjacent blank sheets. Full-size masters that can be filled in electronically and are suitable for projection are available on FOSSweb. Look to the chapter in *Teacher Resources* called Science Notebooks in Middle School for more details on how to use notebooks with FOSS.

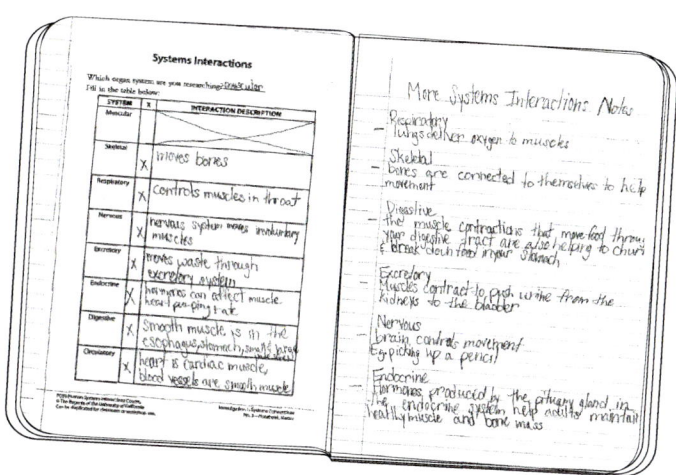

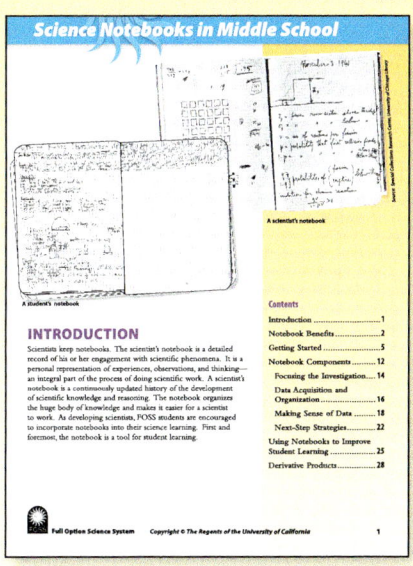

Human Systems Interactions Course—FOSS Next Generation

11

HUMAN SYSTEMS INTERACTIONS — Overview

Reading in Science Resources

Reading is a vital component of the FOSS Program. Reading enhances and extends information and concepts acquired through direct experience.

Readings are included in the **FOSS Science Resources: Human Systems Interactions** book. Students read articles as well as accessing data and information for use in investigations.

Some readings can be assigned as homework or extension activities, whereas other readings have been deemed important for all students to complete with a teacher's support in class.

Each in-class reading has a reading guide embedded in Guiding the Investigation. The reading guide suggests breakpoints with questions to help students connect the reading to their experiences from class, and recommends notebook entries. Each of these readings also includes one or more prompts that ask students to make additional notebook entries. These prompts should help students who missed the in-class reading to process the article in a more meaningful way. Some of the most essential articles are provided as notebook masters. Students can highlight the article as they read, add notes or questions, and add the article to their science notebooks.

Integrating Technology through FOSSweb

The simulations and online activities on FOSSweb are designed to support students' learning at specific times during instruction. Digital resources include streaming videos that can be viewed by the class or small groups.

The FOSSweb and Technology chapter provides details about the online activities for students and the tools and resources for teachers to support and enrich instruction. There are many ways for students to engage with the digital resources—in class as individuals, in small groups, or as a whole class, and at home with family and friends.

FOSS Instructional Design

Assessing Progress

The FOSS assessment system includes both formative and summative assessments. Formative assessment monitors learning during the process of instruction. It measures progress, provides information about learning, and is predominantly diagnostic. Summative assessment looks at the learning after instruction is completed, and it measures achievement.

Formative assessment in FOSS, called **embedded assessment**, is an integral part of instruction, and occurs on a daily basis. You observe action during class in a performance assessment or review notebooks after class. Performance assessments look at students' engagement in science and engineering practices or their recognition of crosscutting concepts. Embedded assessment provides continuous monitoring of students' learning and helps you make decisions about whether to review, extend, or move on to the next idea to be covered.

Benchmark assessments are short summative assessments given after each investigation. These **I-Checks** are actually hybrid tools: they provide summative information about students' achievement, and because they occur soon after teaching each investigation, they can be used diagnostically as well. Reviewing specific items on an I-Check with the class provides additional opportunities for students to clarify their thinking.

The embedded assessments are based on authentic work produced by students during the course of participating in the FOSS activities. Students do their science, and you look at their notebook entries. Bullet points in Guiding the Investigation tell you specifically what students should know and be able to communicate.

If student work is incorrect or incomplete, you know that there has been a breakdown in learning or communications. The assessment system provides a menu of next-step strategies to resolve the situation. Embedded assessment is assessment *for* learning, not assessment *of* learning.

Assessment *of* learning is the domain of the benchmark assessments. Benchmark assessments are delivered at the beginning of the module (entry-level survey) and at the end of the module (posttest), and after each investigation (I-Checks). The benchmark tools are carefully crafted and thoroughly tested assessments composed of valid and reliable items. The assessment items do not simply identify whether a student knows a piece of science content. They also identify the depth to which students understand science concepts and principles and the extent to which they can apply that understanding.

▶ **NOTE**
FOSSmap for teachers and online assessment for students are the technology components of the FOSS assessment system. Students can take assessments online. FOSSmap provides the tools for you to review those assessments online so you can determine next steps for the class or differentiated instruction for individual students based on assessment performance. See the Assessment chapter for more information on these components.

Human Systems Interactions Course—FOSS Next Generation

HUMAN SYSTEMS INTERACTIONS — *Overview*

Solving Real-World Problems

FOSS investigations introduce science content in the context of real-world applications, so that students develop an understanding of how scientific principles explain natural phenomena. By middle school, students can begin to apply this understanding of science to develop solutions to real-world problems.

At this grade level, we ask students to consider problem-solving and engineering challenges that are precise in scope, giving students a thorough understanding of the problem and potential solutions. Students have clear criteria and constraints (in the case of engineering design challenges) and focused topics of research (in the case of research projects).

In the life science, students explore local environments, issues of biodiversity, medical technology applications, and human impact upon ecosystems. In earth science, students consider natural resource supplies and demands, technological advances in space exploration, and human impact affecting Earth's ocean and atmosphere. In physical science, students apply concepts of motion, kinetic energy, heat, and energy transfer in a series of engineering challenges where students develop and refine designs to solve an engineering problem.

Throughout all content areas, students have opportunities to collaborate and develop or select solutions to real-world issues. As described in the NRC Framework, 2011, " . . . engineering and technology provide a context in which students can test their own developing scientific knowledge and apply it to practical problems; doing so enhances their understanding of science—and, for many, their interest in science—as they recognize the interplay among science, engineering, and technology." By providing students with ongoing opportunities to understand and engage with the application of science, we help students develop an appreciation of and enthusiasm for science.

FOSS Instructional Design

Science-Centered Language Development and Standards for Literacy in Science

The FOSS active investigations, science notebooks, *FOSS Science Resources* articles, and formative assessments provide rich contexts in which students develop and exercise thinking and communication. These elements are essential for effective instruction in both science and language arts—students experience the natural world in real and authentic ways and use language to inquire, process information, and communicate their thinking about scientific phenomena. FOSS refers to this development of language process and skills within the context of science as science-centered language development.

In the Science-Centered Language Development in Middle School chapter in *Teacher Resources*, we explore the intersection of science and language and the implications for effective science teaching and language development. Language plays two crucial roles in science learning: (1) it facilitates the communication of conceptual and procedural knowledge, questions, and propositions, and (2) it mediates thinking—a process necessary for understanding. Science provides a real and engaging context for developing literacy, and language-arts skills and strategies to support conceptual development and scientific practices. The skills and strategies used for enhancing reading comprehension, writing expository text, and exercising oral discourse are applied when students are recording their observations, making sense of science content, and communicating their ideas.

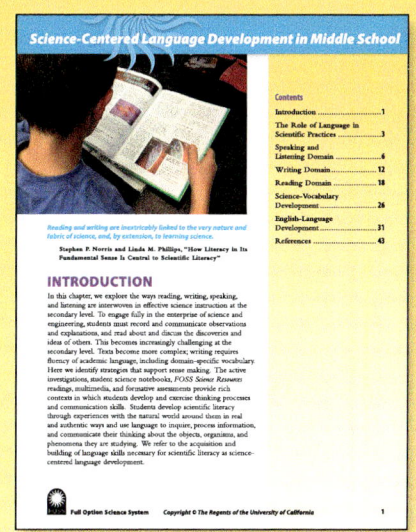

The most effective integration depends on the type of investigation, the experience of students, the language skills and needs of students, and the language objectives that you deem important at the time. The Science-Centered Language Development chapter is a library of resources and strategies for you to use. The chapter describes how literacy strategies are integrated purposefully into the FOSS investigations, gives suggestions for additional literacy strategies that both enhance students' learning in science and develop or exercise English-language literacy skills, and develops science vocabulary with scaffolding strategies for supporting all learners. We identify effective practices in language-arts instruction that support science learning and examine how learning science content and engaging in science and engineering practices support language development.

Specific methods to make connections to the Common Core State Standards for Literacy in Science are included in the flow of Guiding the Investigation. These recommended methods are linked through ELA Connection notes.

Human Systems Interactions Course—FOSS Next Generation

DIFFERENTIATED INSTRUCTION

The roots of FOSS extend back to the mid-1970s and the Science Activities for the Visually Impaired and Science Enrichment for Learners with Physical Handicaps projects (SAVI/SELPH). As those special-education science programs expanded into fully integrated settings in the 1980s, hands-on science proved to be a powerful medium for bringing all students together. The subject matter is universally interesting, and the joy and satisfaction of discovery are shared by everyone. Active science by itself provides part of the solution to full inclusion and provides many opportunities at one time for differentiated instruction.

Many years later, FOSS began a collaboration with educators and researchers at the Center for Applied Special Technology (CAST), where principles of Universal Design for Learning (UDL) had been developed and applied. FOSS continues to learn from our colleagues about ways to use new media and technologies to improve instruction. Here are the UDL principles.

Principle 1. Provide multiple means of representation. Give learners various ways to acquire information and knowledge.

Principle 2. Provide multiple means of action and expression. Offer students alternatives for demonstrating what they know.

Principle 3. Provide multiple means of engagement. Help learners get interested, be challenged, and stay motivated.

The FOSS Program has been designed to maximize the science learning opportunities for students with special needs and students from culturally and linguistically diverse origins. FOSS is rooted in a 35-year tradition of multisensory science education and informed by recent research on UDL. Procedures found effective with students with special needs and students who are learning English are incorporated into the materials and strategies used with all students.

FOSS instruction allows students to express their understanding through a variety of modalities. Each student has multiple opportunities to demonstrate his or her strengths and needs. The challenge is then to provide appropriate follow-up experiences for each student. For some students, appropriate experience might mean more time with the active investigations or online activities. For other students, it might mean more experience building explanations of the science concepts orally or in writing or drawing. For some students, it might mean making vocabulary more explicit through new concrete experiences or through reading to students. For some students, it may be scaffolding

Differentiated Instruction

their thinking through graphic organizers. For other students, it might be designing individual projects or small-group investigations. For some students, it might be more opportunities for experiencing science outside the classroom in more natural, outdoor environments.

The next-step strategies used during the self-assessment sessions after I-Checks provide many opportunities for differentiated instruction. For more on next-step strategies, see the Assessment chapter.

There are additional strategies for providing differentiated instruction. The FOSS Program provides tools and strategies so that you know what students are thinking throughout the module. Based on that knowledge, read through the extension activities for experiences that might be appropriate for students who need additional practice with the basic concepts as well as those ready for more advanced projects. Interdisciplinary extensions are listed at the end of each investigation. Use these ideas to meet the individual needs and interests of your students. In addition, online activities including tutorials and virtual investigations are effective tools to provide differentiated instruction.

English Learners

The FOSS multisensory program provides a rich laboratory for language development for English learners. The program uses a variety of techniques to make science concepts clear and concrete, including modeling, visuals, and active investigations in small groups at centers. Key vocabulary is usually developed within an activity context with frequent opportunities for interaction and discussion between teacher and student and among students. This provides practice and application of the new vocabulary. Instruction is guided and scaffolded through carefully designed lesson plans, and students are supported throughout. The learning is active and engaging for all students, including English learners.

Science vocabulary is introduced in authentic contexts while students engage in active learning. Strategies for helping all students read, write, speak, and listen are described in the Science-Centered Language Development chapter. There is a section on science-vocabulary development with scaffolding strategies for supporting English learners. These strategies are essential for English learners, and they are good teaching strategies for all learners.

Human Systems Interactions Course—FOSS Next Generation

HUMAN SYSTEMS INTERACTIONS — *Overview*

FOCUS QUESTION
How do messages travel to and from the brain?

SCIENCE AND ENGINEERING PRACTICES
Developing and using models

DISCIPLINARY CORE IDEAS
LS1.D: Information processing

CROSSCUTTING CONCEPTS
Structure and function

TEACHING NOTE

This focus question can be answered with a simple yes or no, but the question has power when students support their answers with evidence. Their answers should take the form "Yes, because ____ ."

FOSS INVESTIGATION ORGANIZATION

Courses are subdivided into **investigations** (three in this course). Investigations are further subdivided into two to four **parts**. Each part of each investigation is driven by a **focus question**. The focus question, usually presented as the part begins, signals the challenge to be met, mystery to be solved, or principle to be uncovered. The focus question guides students' actions and thinking and makes the learning goal of each part explicit for teachers over several class sessions. Each part concludes with students recording an answer to the focus question in their notebooks.

The investigation is summarized for the teacher in the At a Glance chart at the beginning of each investigation.

Investigation-specific **scientific background** information for the teacher is presented in each investigation chapter organized by the focus questions.

The **Why Do I Have to Learn This?** section makes direct connections to the NGSS foundation boxes for the grade level—Disciplinary Core Ideas, Science and Engineering Practices, and Crosscutting Concepts. This information is later presented in color-coded sidebar notes to identify specific places in the flow of the investigation where connections to the three dimensions of science learning appear. The section ends with information about teaching and learning and a conceptual-flow graphic of the content.

The **Materials** and **Getting Ready** sections provide scheduling information and detail exactly how to prepare the materials and resources for conducting the investigation. The **Quick Start** table lists of types of planning and preparation.

Teaching Notes and **ELA Connections** appear in blue boxes in the sidebars. These notes compose a second voice in the curriculum—an educative element. The first (traditional) voice is the message you deliver to students. The second educative voice, shared as a teaching note, is designed to help you understand the science content and pedagogical rationale at work behind the instructional scene. ELA Connection boxes provide connections to the Common Core State Standards for English Language Arts.

The **Getting Ready** and **Guiding the Investigation** sections have several features that are flagged in the sidebars. These include several icons to remind you when a particular pedagogical method is suggested, as well as concise bits of information in several categories.

FOSS Investigation Organization

The **safety** icon alerts you to potential safety issues related to chemicals, allergic reactions, and the use of safety goggles.

The small-group **discussion** icon asks you to pause while students discuss data or construct explanations in their groups.

The **vocabulary** icon indicates where students should review recently introduced vocabulary.

The **recording** icon points out where students should make a science-notebook entry.

The **reading** icon signals when the class should read a specific article in the *FOSS Science Resources* book.

The **technology** icon signals when the class should use a digital resource on FOSSweb.

The **assessment** icons appear when there is an opportunity to assess student progress by using embedded or benchmark assessments. Some are performance assessments—observations of science and engineering practices, crosscutting concepts, and core ideas, indicated by an icon that includes a beaker and ruler.

The **engineering** icon indicates opportunities for an experience incorporating engineering practices.

The **math** icon indicates an opportunity to engage in numerical data analysis and mathematics practice.

The **homework** icon indicates science learning experiences that extend beyond the classroom.

The **EL note** provides a specific strategy to assist English learners in developing science concepts.

EL NOTE

To help with scheduling, you will see icons for the start of a new **session** within an investigation part.

SESSION 2 *45–50 minutes*

HUMAN SYSTEMS INTERACTIONS — *Overview*

MANAGEMENT STRATEGIES

FOSS has tried to anticipate the most likely learning environments in which science will be taught and designed the curriculum to be effective in those settings. The most common setting is the 1-hour period (45–55 minutes) every day, one teacher, in the science room. Students come in wave after wave, and they all learn the same thing. Some teachers may have two preps because they teach seventh-grade and eighth-grade classes. The **Human Systems Interactions Course** was designed to work effectively in this traditional hour-a-day format.

The 1-hour subdivisions of the course adapt nicely to the block-scheduling model. It is usually possible to conduct two of the 1-hour sessions in a 90-minute block because of the uninterrupted instructional period. A block allows students to set up an experiment and collect, organize, and process the data all in one sequence. Block scheduling is great for FOSS; students learn more, and teachers are responsible for fewer preps.

Interdisciplinary teams of teachers provide even more learning opportunities. Students will be using mathematics frequently and in complex ways to extract meaning from their inquiries. It has been our experience, however, that middle school students are not skilled at applying mathematics in science because they have had few opportunities to use these skills in context. In an interdisciplinary team, the math teacher can use student-generated data to teach and enhance math skills and application.

The integration of other subject areas, such as language arts, into the science curriculum is also enhanced when interdisciplinary teams are used.

Managing Time

Time is a precious commodity. It must be managed wisely in order to realize the full potential of your FOSS curriculum. The right amount of time should be allocated for preparation, instruction, discussion, assessment, research, and current events. Start from the premise that there will not be enough time to do everything, so you will have to budget selectively. Don't scrimp on the prep time, particularly the first time you use the curriculum. Spend enough time with the *Investigations Guide* to become completely familiar with the lesson plans. Take extra time at the start of the course to set up your space efficiently; you will be repaid many times over later. As you become more familiar with the FOSS Program and the handling of the materials, the proportion of time devoted to each aspect of the program may shift, so that you are spending more and more time on instruction and enrichment activities.

Management Strategies

Effective use of time during the instructional period is one of the keys to a great experience with this course. The *Investigations Guide* offers suggestions for keeping the activities moving along at a good pace, but our proposed timing will rarely exactly match yours. The best way we know for getting in stride with the curriculum is to start teaching it. Soon you will be able to judge where to break an activity or push in a little enrichment to fill your instructional period.

Managing Space

The **Human Systems Interactions Course** will work in the ideal setting: flat-topped tables where students work with materials in groups of four; theater seating for viewing online activities (darkened); technology available for accessing FOSSweb on the Internet for online activities, videos, and references. But we don't expect many teachers to have the privilege of working in such a space. So we designed FOSS courses to work effectively in a number of typical settings, including the science lab and regular classroom. We have described, however, the minimum space and resources needed to use FOSS. Here's the list, in order of importance.

- A computer with Internet access, and a large-screen display monitor or projector
- Tables or desks for students to work in groups of four
- A whiteboard, blackboard, or chart paper and marking pens
- A surface for materials distribution
- A place to clean and organize equipment
- A convenient place to store the kit
- A computer lab or multiple digital devices

Once the minimum resources are at hand, take a little time to set up your science area. This investment will pay handsome dividends later since everyone will be familiar with the learning setup.

- Organize your computer and projection system and be sure the Internet connection is working smoothly.
- Think about the best organization of furniture. This may change from investigation to investigation.
- Plan where to set up your materials stations.
- Know how students will keep notes and record data, and plan where students will keep their notebooks.

HUMAN SYSTEMS INTERACTIONS — *Overview*

Managing Students

A typical class of middle school students is a wonderfully complex collection of personalities, including the clown, the athlete, the fashion statement, the worrier, the achiever, the pencil sharpener, the show-off, the reader, and the question-answerer. Notice there is no mention of the astrophysicist, but she could be in there, too. Management requires delicate coordination and flexibility—some days students take their places in an orderly fashion and sit up straight in their chairs, fully prepared to learn. Later in the week, they are just as likely to have the appearance of migrating waterfowl, unable to find their place, talkative, and constantly moving.

FOSS employs a number of strategies for managing students. Often a warm-up activity is a suitable transition from lunch or the excitement of changing rooms to the focused intellectual activities of the **Human Systems Interactions Course**. Warm-ups tend to be individual exercises that review what transpired yesterday with a segue to the next development in the curriculum. This gives students time to get out their notebooks, grind points on their pencils, settle into their space, and focus.

Students most often work in groups in this course. Groups of four are generally used, but at other times, students work in pairs.

Suggestions for guiding students' work in collaborative groups are described later in this chapter.

When Students Are Absent

When a student is absent for a session, another student can act as a peer tutor and share the science notebook entries made for that day. The science notebooks should be a valuable tool for students to share in order to catch up on missed classes. Also consider giving him or her a chance to spend some time with the materials.

Students can use the resources on FOSSweb at school or at home for the missed class. And finally, allow the student to bring home *FOSS Science Resources* to read any relevant articles. Each article has a few review items that the student can respond to verbally or in writing.

Management Strategies

Managing Technology

The **Human Systems Interactions Course** includes an online component. The online activities and materials are not optional. For this reason, it is essential that you have in your classroom at minimum one computer, a large-screen display monitor or projection system, and a connection to the Internet. Sometimes you will use multimedia to make presentations to the entire class. Sometimes small groups or individuals will use the online program to work simulations and representations, and to gather information. Plan on the students having access to computers or tablets for work in groups for these sessions.

- Investigation 1, Parts 1 and 2
- Investigation 2, Parts 1 and 2
- Investigation 3, Parts 1 and 3

Option 1: The computer lab. If you have access to a lab where all students can work simultaneously as individuals, pairs, or small groups, schedule time in the lab for your classes. If you have access to a cart with a class set of devices, schedule that for your classroom.

Option 2: Classroom computers or other digital devices. With multiple devices for groups in the science classroom, you can set up a multitasking environment with half the students working with Internet resources and half engaged in reading or small-group discussions. Then swap roles. If every student or pair has access to a device, you are all set.

Option 3: Home access. Students can access FOSSweb from home by visiting www.FOSSweb.com and accessing the class pages with the account information you provide for student use. You must set up a class page for students to have home access to the multimedia.

HUMAN SYSTEMS INTERACTIONS — *Overview*

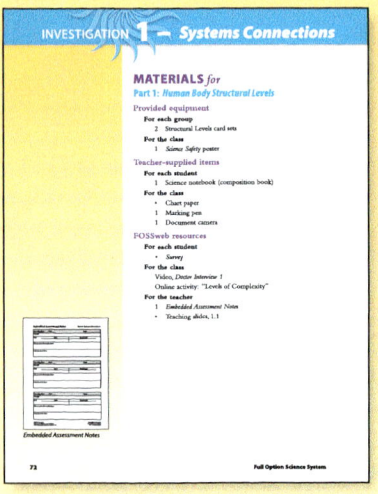

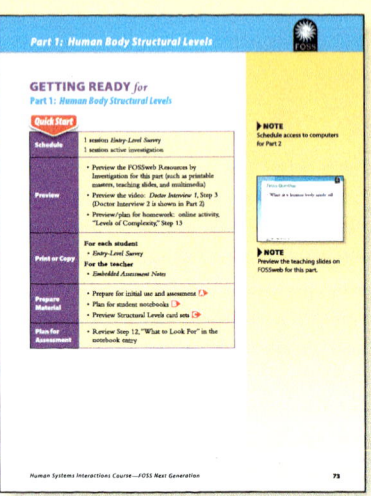

Managing Materials

The Materials section lists the items in the equipment kit and any teacher-supplied materials. It also describes things to do to prepare a new kit and how to check and prepare the kit for your classroom. Individual photos of each piece of FOSS equipment are available for printing from FOSSweb, and can help students and you identify each item.

The FOSS Program designers suggest using a central materials distribution system. You organize all the materials for an investigation at a single location called the materials station. As the investigation progresses, one member of each group gets materials as they are needed, and another returns the materials when the investigation is complete. You place the equipment and resources at the station, and students do the rest. Students can also be involved in cleaning and organizing the materials at the end of a session.

The Materials list for each investigation is divided into these categories.

- Provided equipment found in the FOSS equipment kit
- Teacher-supplied items
- FOSSweb resources to be downloaded or projected

Each category is further subdivided by need.

- For each student
- For each group
- For the class
- For the teacher

The Getting Ready section begins with the Quick Start in a table to help the teacher immediately know the schedule, what to preview, print or copy, prepare in terms of materials, and what to plan for assessment. Brief preparation details linked to the Quick Start provide specific information.

WORKING IN COLLABORATIVE GROUPS

Collaboration is important in science. Scientists usually collaborate on research enterprises. Groups of researchers often contribute to the collection of data, the analysis of findings, and the preparation of the results for publication.

Collaboration is expected in the science classroom, too. Some tasks call for everyone to have the same experience, either taking turns or doing the same things simultaneously. At other times, group members may have different experiences that they later bring together.

Research has shown that students learn better and are more successful when they collaborate. Working together promotes student interest, participation, learning, and self-confidence. FOSS investigations use collaborative groups extensively.

No single model for collaborative learning is promoted by FOSS. We can suggest, however, a few general guidelines that have proven successful over the years.

For most activities in middle school, collaborative groups of four in which students take turns assuming specific responsibilities work best. Groups can be identified completely randomly (first four names drawn from a hat constitute group 1), or you can assemble groups to ensure diversity. Thoughtfully constituted groups tend to work better.

Groups can be maintained for extended periods of time, or they can be reconfigured more frequently. For a short course, you might keep students in the same groups for the entire course.

Functional roles within groups can be determined by the members themselves, or they can be assigned in one of several ways. Each member in a collaborative group can be assigned a number or a color. Then you need only announce which color or number will perform a certain task for the group at a certain time. Compass points can also be used: the person seated on the east side of the table will be the Reporter for this investigation.

The functional roles used in the investigations follow. If you already use other names for functional roles in your class, use those in place of these in the investigations.

Human Systems Interactions Course—FOSS Next Generation

HUMAN SYSTEMS INTERACTIONS — *Overview*

Getters are responsible for materials. One person from each group gets equipment from the materials station, and another person later returns the equipment.

One person is the **Starter** for each task. This person makes sure that everyone gets a turn and that everyone has an opportunity to contribute ideas to the investigation.

The **Recorder** collects data as it happens and makes sure that everyone has recorded information on his or her science notebook sheets.

The **Reporter** shares group data with the class or transcribes it to the board or class chart.

Getting started with collaborative groups requires patience, but the rewards are great. Once collaborative groups are in place, you will be able to engage students more in meaningful conversations about science content. You are free to "cruise" the groups, to observe and listen to students as they work, and to interact with individuals and small groups as needed.

SAFETY IN THE CLASSROOM

Following the procedures described in each investigation will make for a very safe experience in the classroom. You should also review your district safety guidelines and make sure that everything that you do is consistent with those guidelines. The *FOSS Safety* poster is included in the kit for classroom use, and the safey guidelines are in the *FOSS Science Resources* book for student reference.

Look for the safety icon in the Getting Ready and Guiding the Investigation sections, which will alert you to safety considerations throughout the course.

Safety Data Sheets (SDS) for materials used in the FOSS Program can be found on FOSSweb. If you have questions regarding any SDS, call Delta Education at 1-800-258-1302 (Monday–Friday 8 a.m. to 5 p.m. ET).

General classroom safety rules to share with students are listed here.

1. Always follow the safety procedures outlined by your teacher. Follow directions, and ask questions if you're unsure of what to do.
2. Never put any material in your mouth. Do not taste any material or chemical unless your teacher specifically tells you to do so.
3. Do not smell any unknown material. If your teacher tells you to smell a material, wave a hand over it to bring the scent toward your nose.
4. Avoid touching your face, mouth, ears, eyes, or nose while working with chemicals, plants, or animals. Tell your teacher if you have any allergies.
5. Always wash your hands with soap and warm water immediately after using chemicals (including common chemicals, such as salt and dyes) and handling natural materials or organisms.
6. Do not mix unknown chemicals just to see what might happen.
7. Always wear safety goggles when working with liquids, chemicals, and sharp or pointed tools. Tell your teacher if you wear contact lenses.
8. Clean up spills immediately. Report all spills, accidents, and injuries to your teacher.
9. Treat animals with respect, caution, and consideration.
10. Never use the mirror of a microscope to reflect direct sunlight. The bright light can cause permanent eye damage.

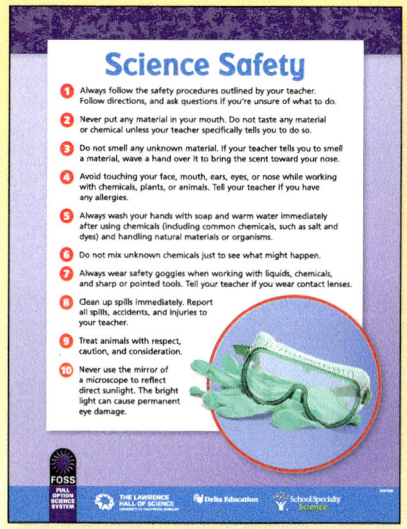

HUMAN SYSTEMS INTERACTIONS — *Overview*

FOSS CONTACTS

General FOSS Program Information

www.FOSSweb.com

www.DeltaEducation.com/FOSS

Developers at the Lawrence Hall of Science

foss@berkeley.edu

Customer Service at Delta Education

http://www.DeltaEducation.com/contact.aspx

Phone: 1-800-258-1302, 8:00 a.m.–5:00 p.m. ET

FOSSmap (online component of FOSS assessment system)

http://fossmap.com/

FOSSweb account questions/help logging in

School Specialty Online Support

loginhelp@schoolspecialty.com

Phone: 1-800-513-2465, 8:30 a.m.–6:00 p.m. ET
5:30 a.m.–3:00 p.m. PT

FOSSweb Tech Support

support@fossweb.com

Professional development

http://www.FOSSweb.com/Professional-Development

Safety issues

www.DeltaEducation.com/SDS.shtml

Phone: 1-800-258-1302, 8:00 a.m.–5:00 p.m. ET

For chemical emergencies, contact Chemtrec 24 hours a day.

Phone: 1-800-424-9300

Sales and Replacement Parts

www.DeltaEducation.com/foss/buy

Phone: 1-800-338-5270, 8:00 a.m.–5:00 p.m. ET

HUMAN SYSTEMS INTERACTIONS — *Framework and NGSS*

INTRODUCTION

This chapter provides details on how this FOSS middle school course fits into the larger matrix of the FOSS Program. Each FOSS module K–5 and middle school course 6–8 was designed from FOSS conceptual frameworks that were developed based on a decade of research on the FOSS Program and the influence of *A Framework for K–12 Science Education* (2012) and *Next Generation Science Standards* (NGSS, 2013).

The FOSS curriculum provides a coherent vision of science education in three ways as described by the *Framework*. First, FOSS is designed around learning as a development progression, providing experiences that allow students to continually build on their initial notions and develop more complex scientific and engineering ideas. Students develop understanding over time by building on foundational elements or intermediate knowledge. Those elements are detailed in the conceptual frameworks.

Second, FOSS limits the number of core ideas, choosing depth of knowledge over superficial coverage. Those core ideas are addressed at multiple grade levels in ever more complex ways. FOSS investigations at each grade level focus on elements of core ideas that are teachable and learnable at that grade level.

And third, FOSS investigations facilitate the integration of scientific knowledge (content knowledge) with practices of science and engineering by providing students with firsthand experiences.

If this is your first time teaching a FOSS middle school course, you should review this material but save an in-depth study of it until after you have experienced the course in the classroom with students. Teach the course with the confidence that the developers have carefully considered the latest research and have integrated into each investigation the three dimensions of the *Framework* and *NGSS*, and have designed powerful connections to the Common Core State Standards for English Language Arts.

Contents

Introduction	29
FOSS Conceptual Frameworks	30
Background for the Conceptual Framework	32
Connections to NGSS	36
FOSS Next Generation K–8 Scope and Sequence	48

▶ REFERENCES

National Research Council. *A Framework for K–12 Science Education: Practices, Crosscutting Concepts, and Core Ideas.* Washington, DC: The National Academies Press, 2012.

NGSS Lead States. *Next Generation Science Standards: For States, By States.* Washington, DC: The National Academies Press, 2013.

National Governors Association Center for Best Practices and Council of Chief State School Officers. *Common Core State Standards for English Language Arts & Literacy in History/Social Studies, Science, and Technical Subjects*, 2010.

FOSS Full Option Science System

FOSS CONCEPTUAL FRAMEWORKS

FOSS has conceptual structure at the course level. The concepts are carefully selected and organized in a sequence that makes sense to students when presented as intended. In the last half decade, research has been focused on learning progressions. The idea behind a learning progression is that **core ideas** in science are complex and wide-reaching—ideas such as the structure of matter or the relationship between the structure and function of organisms. From the age of awareness throughout life, matter and organisms are important to us. There are things we can and should understand about them in our primary school years, and progressively more complex and sophisticated things we should know about them as we gain experience and develop our cognitive abilities. When we as educators can determine those logical progressions, we can develop meaningful and effective curriculum.

FOSS has elaborated learning progressions for core ideas in science for kindergarten through grade 8. Developing the learning progressions involves identifying successively more sophisticated ways of thinking about core ideas over multiple years. "If mastery of a core idea in a science discipline is the ultimate educational destination, then well-designed learning progressions provide a map of the routes that can be taken to reach that destination" (National Research Council, *A Framework for K–12 Science Education*, 2012).

The FOSS modules (grades K–5) and courses (grades 6–8) are organized into three domains: physical science, earth science, and life science. Each domain is divided into two strands, which represent a core scientific idea, as shown in the columns in the table: matter/energy and change, atmosphere and Earth/rocks and landforms, structure and function/complex systems. The sequence of modules and courses in each strand relates to the core ideas described in the national framework. Modules at the bottom of the table form the foundation in the primary grades. The core ideas develop in complexity as you proceed up the columns.

In addition to the science content framework, every course provides opportunities for students to engage in and understand science practices, and many courses explore issues related to engineering practices and the use of natural resources.

FOSS Conceptual Frameworks

The science content used to develop the FOSS courses describes what we want students to learn; the science and engineering practices describe how we want students to learn. Practices involve a number of habits of mind and philosophical orientations, and these, too, will develop in richness and complexity as students advance through their science studies. Science and engineering practices involve behaviors, so they can be best assessed while in progress. Thus, assessment of practices is based on teacher observation. The indicators of progress include students involved in the many aspects of active thinking, students motivated to learn, and students taking responsibility for their own learning.

FOSS Next Generation—K–8 Sequence

	PHYSICAL SCIENCE		EARTH SCIENCE		LIFE SCIENCE	
	MATTER	ENERGY AND CHANGE	ATMOSPHERE AND EARTH	ROCKS AND LANDFORMS	STRUCTURE/ FUNCTION	COMPLEX SYSTEMS
6–8	Waves; Gravity and Kinetic Energy Chemical Interactions Electromagnetic Force; Variables and Design		Planetary Science Earth History Weather and Water		Heredity and Adaptation Human Systems Interactions Populations and Ecosystems Diversity of Life	
5	Mixtures and Solutions		Earth and Sun		Living Systems	
4		Energy		Soils, Rocks, and Landforms	Environments	
3	Motion and Matter		Water and Climate		Structures of Life	
2	Solids and Liquids			Pebbles, Sand, and Silt	Insects and Plants	
1		Sound and Light	Air and Weather		Plants and Animals	
K	Materials and Motion		Trees and Weather		Animals Two by Two	

Human Systems Interactions Course—FOSS Next Generation

31

HUMAN SYSTEMS INTERACTIONS — Framework and NGS.

BACKGROUND FOR THE CONCEPTUAL FRAMEWORKS

Systems of the Human Body

To appreciate the wonder of being human requires a deep, dedicated systemic point of view. First it is necessary to understand that life happens in cells. You are the result of a mass of several trillion individual cells all working in concert. Whipping those cells into shape to produce you is a masterpiece of coordination. Those trillions of cells have a demanding set of support criteria that must be attended to continuously. No mean feat!

Cells need a steady supply of food, water, and oxygen. Full service at all times. The human system is replete with dozens of subsystems dedicated to servicing cells. Cells "eat" a very limited menu of simple chemical foods (nutrients). Leading the list is glucose, a cell's favorite source of energy. Cell nutrients are extracted from food that we eat. The digestive system is dedicated to the process of extracting nutrient chemicals from the complex mess of organic substances we ingest. Once the nutrients are separated from the other stuff, they diffuse from the digestive system into the blood.

Nutrients move along in the blood as it flows through the circulatory system. The signature feature of the circulatory system is an extensive network of blood vessels. A critically important fixture in the network of vessels is a heart, the durable muscle that pumps the blood to the lungs and then throughout the entire body.

The vessel network includes arteries, large vessels that divide and divide, branching out and getting smaller and smaller as they reach toward the cells in your body. When the arteries near their destinations, they divide one final time into tiny capillaries.

Capillaries are so numerous, small, and delicate that they come into contact with most cells in your body. The thin walls of the capillaries allow cell nutrients to pass from the bloodstream into the cells. The blood then continues on its way. The capillaries converge with one another into larger and larger vessels. These vessels—veins—carry the spent blood back to the heart.

The spent blood in veins eventually enter the right atrium. From there, blood enters the right ventricle, which pumps it to the lungs, where the circulatory system merges with the respiratory system. The blood releases its load of waste carbon dioxide into the lungs and picks up a fresh charge of oxygen. The refreshed blood collects in the left atrium and then moves into the left ventricle, which pumps it out through the aorta into the body.

FOSS Conceptual Frameworks

Round and round the blood goes, pushing about 25 trillion red blood cells on their way through the circulatory system with each cycle. Red blood cells live about 4 months, and then they die and are replaced. Your body manufactures red blood cells at the rate of 2 million per second. That's another subsystem that is tightly coordinated to ensure that the blood always has enough red blood cells to provide oxygen and remove carbon dioxide for all your trillions of cells on a continuous basis.

The circulatory system interacts with other systems to maintain the health of your cells. It works with the respiratory system to exchange gases, and with the digestive system to acquire nutrient chemicals that cells use for energy generation, growth, and structural repair. It also works with the hepatic (liver) system and renal (kidney) system. These two systems act as selective filters for removing specific classes of waste materials from the blood. These wastes are dumped into blood as it passes through the capillaries.

The nervous system is the complex electric system that coordinates and manages all the other systems in a human. It is managed by the central nervous system, which comprises the three parts of the brain (cerebrum, cerebellum, brain stem) and the spinal cord. The peripheral nervous system includes all the millions of receptor neurons that gather information from the environment, which is sent to the brain, and the network of motor neurons that convey action instructions from the brain to muscles and other tissues.

> **CONCEPTUAL FRAMEWORK**
> **Life Science, Focus on Structure and Function:**
> **Human Systems Interactions**
>
> **Structure and Function**
>
> **Concept A** All living things need food, water, a way to dispose of waste, and an environment in which they can live.
> - The cell is the basic unit of life. All organisms are one or more cells.
> - Aerobic cellular respiration is the process by which energy stored in food molecules is converted into usable energy for cells.
> - In multicellular organisms, such as humans, cells form tissues, tissues form organs, and organs form organ systems (subsystems), which interact to serve the needs of the organism.
> - The human body is a system of interacting subsystems (circulatory, digestive, endocrine, excretory, muscular, nervous, respiratory, skeletal).
>
> **Concept C** Animals detect, process, and use information about their environment to survive.
> - The nervous system is a human subsystem that functions to gather and synthesize information from the environment.
> - Sensory receptors are structures that respond to stimuli by sending messages to the brain for processing and response.
> - Neural pathways change and grow as information is acquired and stored as memories.
>
> **Complex Systems**
>
> **Concept A** Organisms and populations of organisms depend on their environmental interactions with other living things and with nonliving factors.
> - The Sun provides energy that plants use to produce food molecules from carbon dioxide and water. The energy in food molecules is processed in the cells of most organisms to drive life processes.

Everything that you do is managed by your nervous system. You can breathe and maintain a constant heart function even while sound asleep. You have neurons distributed throughout your body that monitor these and other functions, sending instructions to the proper places to keep your heart beating and your diaphragm pulling air into your lungs.

Human Systems Interactions Course—FOSS Next Generation 33

HUMAN SYSTEMS INTERACTIONS — Framework and NGSS

Life Science Content Sequence

This table shows all the modules and courses grades 3–8 in the FOSS content sequence for Life Science with an emphasis on the modules/courses that inform the structure and function strand. The supporting elements in these modules (somewhat abbreviated) are listed. The elements for the **Human Systems Interactions** are expanded to show how they fit into the sequence.

Module or course	LIFE SCIENCE — Structure and Function	Complex Systems
Human Systems Interactions (middle grades)		
Diversity of Life (middle grades)	• All living things are made of cells (unicellular or multicellular). Special structures within cells are responsible for various functions. • Cells have the same needs and perform the same functions as more complex organisms. • All living things need food, water, a way to dispose of waste, and an environment in which they can live (macro and micro levels). • Plants reproduce in a variety of ways, sometimes depending on animal behaviors and specialized features for reproduction.	• Adaptations are structures or behaviors of organisms that enhance their chances to survive and reproduce in their environment. • Biodiversity is the wide range of existing life-forms that have adapted to the variety of conditions on Earth, from terrestrial to marine ecosystems.
Living Systems (grade 5)	• Food is digested to provide animals with the materials they need for body repair and growth and to release the energy they need to maintain body warmth and for motion. • Reproduction is essential to the continued existence of every kind of organism. • Humans and other animals have systems made up of organs that are specialized for particular body functions. • Animals detect, process, and use information about their environment to survive.	• Organisms obtain gases, water, and minerals from the environment and release waste matter back into the environment. • Matter cycles between air and soil, and among plants, animals, and microbes as these organisms live and die. • Organisms are related in food webs. • Some organisms, such as fungi and bacteria, break down dead organisms, operating as decomposers.
Structures of Life (grade 3)	• A seed is a living organism. • Plants and animals have structures that function in growth, survival, and reproduction. • Reproduction is essential to the continued existence of every kind of organism. • Plants and animals grow and change and have predictable characteristics at different stages. • Bones have several functions: support, protection, and movement.	• Organisms are related in food chains. • Animals exhibit different kinds of behaviors. • Different organisms can live in different environments; organisms have adaptations that allow them to survive in that environment. • Changes in an organism's habitat are sometimes beneficial to it and sometimes harmful. • A skeleton is a system of interacting bones. The skeletons of humans and other mammals have many similarities.
Plants and Animals (grade 1)	• Plants and animals have structures, and animals have behaviors that help the organisms grow and survive in their habitat. • Seeds and bulbs are alive. • Plants need water, light, air, and space. • Plants don't live forever. New plants can grow from seeds, bulbs, roots, and stems.	• Plants make their own food. • Animals eat plants and other animals. • A habitat is a place where plants and animals live. There are many different kinds of habitats.

Full Option Science System

FOSS Conceptual Framework

Human Systems Interactions

Structure and Function	Complex Systems
• The cell is the basic unit of life. All living things are one or more cells. • Aerobic cellular respiration is the process by which energy stored in food molecules is converted into usable energy for cells. • In multicellular organisms, such as humans, cells form tissues, tissues form organs, and organs form organ systems (subsystems), which interact to serve the needs of the organism. • The human body is a system of interacting subsystems (circulatory, digestive, endocrine, excretory, muscular, nervous, respiratory, skeletal). • The nervous system is a human subsystem that functions to gather and synthesize information from the environment. • Sensory receptors are structures that respond to stimuli by sending messages to the brain for processing and response. • Neural pathways change and grow as information is acquired and stored as memories.	• The Sun provides energy that plants use to produce food molecules from carbon dioxide and water. The energy in food molecules is processed in the cells of most organisms to drive life processes.

▶ **NOTE**
See the Assessment chapter at the end of this *Investigations Guide* for more details on how the FOSS embedded and benchmark assessment opportunities align to the conceptual frameworks and the learning progressions. In addition, the Assessment chapter describes specific connections between the FOSS assessments and the NGSS performance expectations.

The NGSS Performance Expectations addressed in this course include:

Life Sciences
MS-LS1-1
MS-LS1-3
MS-LS1-7
MS-LS1-8

See pages 46–47 in this chapter for more details on the Grades 6–8 NGSS Performance Expectations.

Human Systems Interactions Course—FOSS Next Generation

HUMAN SYSTEMS INTERACTIONS — Framework and NGSS

CONNECTIONS TO NGSS

Science and Engineering Practices	Connections to Common Core State Standards—ELA
Asking questions Analyzing and interpreting data Constructing explanations Engaging in argument from evidence Obtaining, evaluating, and communicating information	**Reading—Literacy in Science and Technical Subjects** 1. Cite specific textual evidence to support analysis of science and technical texts. 4. Determine the meaning of symbols, key terms, and other domain-specific words and phrases as they are used in a specific scientific or technical context relevant to grades 6–8 texts and topics. 5. Analyze the structure an author uses to organize a text, including how the major sections contribute to the whole and to an understanding of the topic. 7. Integrate quantitative or technical information expressed in words in a text with a version of that information expressed visually (e.g., in a flowchart, diagram, model, graph, or table). 10. Read and comprehend science/technical texts in the grades 6–8 text complexity band independently and proficiently. **Writing—Literacy in Science and Technical Subjects** 8. Gather relevant information from multiple print and digital sources, using search terms effectively. **Speaking and Listening** 1. Engage effectively in a range of collaborative discussions (one-on-one, in groups, and teacher led) with diverse partners on middle school topics, texts, and issues, building on others' ideas and expressing their own clearly. 3. Delineate a speaker's argument and specific claims, evaluating the soundness of the reasoning and relevance and sufficiency of the evidence and identifying when irrelevant evidence is introduced. 4. Present claims and findings, emphasizing salient points in a focused, coherent manner with relevant evidence, sound valid reasoning, and well-chosen details; use appropriate eye contact, adequate volume, and clear pronunciation. **Language** 4. Determine or clarify the meaning of unknown words or phrases. 4b. Use Greek or Latin affixes and roots as clues to the meaning of a word.

Inv. 1: Systems Connections

Connections to NGSS

Disciplinary Core Ideas

LS1.A: Structure and function
- All living things are made up of cells, which is the smallest unit that can be said to be alive. An organism may consist of one single cell (unicellular) or many different numbers and types of cells (multicellular).
- In multicellular organisms, the body is a system of multiple interacting subsystems. These subsystems are groups of cells that work together to form tissues and organs that are specialized for particular body functions.

Crosscutting Concepts

Cause and effect
Scale, proportion, and quantity
Systems and system models
Structure and function

HUMAN SYSTEMS INTERACTIONS — *Framework and NGS*

Inv. 2: Supporting Cells

Science and Engineering Practices

Developing and using models
Constructing explanations
Obtaining, evaluating, and communicating information

Connections to Common Core State Standards—ELA

Reading—Literacy in Science and Technical Subjects
1. Cite specific textual evidence to support analysis of science and technical texts.
2. Determine the central ideas or conclusions of a text; provide an accurate summary of the text distinct from prior knowledge or opinions.
4. Determine the meaning of symbols, key terms, and other domain-specific words and phrases as they are used in a specific scientific or technical context relevant to grades 6–8 texts and topics.
6. Analyze the author's purpose in providing an explanation, describing a procedure, or discussing an experiment in a text.
7. Integrate quantitative or technical information expressed in words in a text with a version of that information expressed visually (e.g., in a flowchart, diagram, model, graph, or table).

Writing—Literacy in Science and Technical Subjects
5. With some guidance and support from peers and adults, develop and strengthen writing as needed by planning, revising, editing, rewriting, or trying a new approach, focusing on how well purpose and audience have been addressed.
7. Conduct short research projects to answer a question (including a self-generated question), drawing on several sources and generating additional related, focused questions that allow for multiple avenues of exploration.
8. Gather relevant information from multiple print and digital sources, using search terms effectively.

Speaking and Listening
1. Engage effectively in a range of collaborative discussions (one-on-one, in groups, and teacher led) with diverse partners on middle school topics, texts, and issues, building on others' ideas and expressing their own clearly.

Language
5. Demonstrate understanding of word relationships and nuances in word meaning.
6. Acquire and use academic and domain-specific words and phrases.

Connections to NGSS

Disciplinary Core Ideas		Crosscutting Concepts
LS1.A: Structure and function • In multicellular organisms, the body is a system of multiple interacting subsystems. These subsystems are groups of cells that work together to form tissues and organs that are specialized for particular body functions. **LS1.C: Organization for matter and energy flow in organisms** • Within individual organisms, food moves through a series of chemical reactions in which it is broken down and rearranged to form new molecules, to support growth, or to release energy. **PS3.D: Energy in chemical processes and everyday life** • Cellular respiration in plants and animals involves chemical reactions with oxygen that release stored energy. In these processes, complex molecules containing carbon react with oxygen to produce carbon dioxide and other materials.		Scale, proportion, and quantity Systems and system models Energy and matter

HUMAN SYSTEMS INTERACTIONS — *Framework and NGS*

Inv. 3: The Nervous System

Science and Engineering Practices

Developing and using models
Planning and carrying out investigations
Analyzing and interpreting data
Constructing explanations
Engaging in argument from evidence
Obtaining, evaluating, and communicating information

Connections to Common Core State Standards—ELA

Reading—Literacy in Science and Technical Subjects
1. Cite specific textual evidence to support analysis of science and technical texts.
2. Determine the central ideas or conclusions of a text; provide an accurate summary of the text distinct from prior knowledge or opinions.
6. Analyze the author's purpose in providing an explanation, describing a procedure, or discussing an experiment in a text.
7. Integrate quantitative or technical information expressed in words in a text with a version of that information expressed visually (e.g., in a flowchart, diagram, model, graph, or table).
9. Compare and contrast the information gained from experiments, simulations, video, or multimedia sources with that gained from reading a text on the same topic.
10. Read and comprehend science/technical texts in the grades 6–8 text complexity band independently and proficiently.

Writing—Literacy in Science and Technical Subjects
7. Conduct short research projects to answer a question, drawing on several sources and generating additional related, focused questions that allow for multiple avenues of exploration.
8. Gather relevant information from multiple print and digital sources, using search terms effectively.
9. Draw evidence from informational texts to support analysis reflection, and research.

Speaking and Listening
1. Engage effectively in a range of collaborative discussions (one-on-one, in groups, and teacher led) with diverse partners on middle school topics, texts, and issues, building on others' ideas and expressing their own clearly.
3. Delineate a speaker's argument and specific claims, evaluating the soundness of the reasoning and relevance and sufficiency of the evidence and identifying when irrelevant evidence is introduced.
6. Adapt speech to a variety of contexts and tasks, demonstrating command of formal English when indicated or appropriate.

Language
4. Determine or clarify the meaning of unknown words or phrases.
6. Acquire and use academic and domain-specific words and phrases.

Connections to NGSS

Disciplinary Core Ideas

LS1.A: Structure and function
- In multicellular organisms, the body is a system of multiple interacting subsystems. These subsystems are groups of cells that work together to form tissues and organs that are specialized for particular body functions.

LS1.D: Information processing
- Each sense receptor responds to different inputs (electromagnetic, mechanical, chemical), transmitting them as signals that travel along nerve cells to the brain. The signals are then processed in the brain, resulting in immediate behaviors or memories.

Crosscutting Concepts

Patterns
Cause and effect
Scale, proportion, and quantity
Systems and system models
Structure and function

HUMAN SYSTEMS INTERACTIONS — Framework and NGSS

SCIENCE AND ENGINEERING PRACTICES

A Framework for K–12 Science Education (National Research Council, 2012) describes eight science and engineering practices as essential elements of a K–12 science and engineering curriculum.

The learning progression for this dimension of the framework is addressed in *Next Generation Science Standards* (National Academies Press, 2013), volume 2, appendix F. Elements of the learning progression for practices recommended for grades 6–8 as described in the performance expectations appear in bullets below each practice.

Science and Engineering Practices Addressed

1. **Asking questions**
   - Identify and/or clarify evidence and/or the premise(s) of an argument.

2. **Developing and using models**
   - Develop and/or use a model to predict and/or describe phenomena.
   - Develop and/or use a model to generate data to test ideas about phenomena in natural or designed systems, including those representing inputs and outputs, and those at unobservable scales.

3. **Planning and carrying out investigations**
   - Collect data to produce data to serve as the basis for evidence to answer scientific questions or test design solutions under a range of conditions.

4. **Analyzing and interpreting data**
   - Analyze and interpret data to provide evidence for phenomena.

5. **Constructing explanations**
   - Apply scientific ideas, principles, and/or evidence to construct, revise, and/or use an explanation for real-world phenomena, examples, or events.

6. **Engaging in argument from evidence**
   - Respectfully provide and receive critiques about one's explanations, procedures, models, and questions by citing relevant evidence and posing and responding to questions that elicit pertinent elaboration and detail.
   - Construct, use, and/or present an oral and written argument supported by empirical evidence and scientific reasoning to support or refute an explanation or a model for a phenomenon (or a solution to a problem).

7. **Obtaining, evaluating, and communicating information**
   - Critically read scientific texts adapted for classroom use to determine the central ideas and/or obtain scientific and/or technical information to describe patterns in and/or evidence about the natural and designed world(s).

Connections to NGSS

- Integrate qualitative and/or quantitative scientific and/or technical information in written text with that contained in media and visual displays to clarify claims and findings.

- Communicate scientific and/or technical information (e.g., about a proposed object, tool, process, system) in writing and/or through oral presentations.

Crosscutting Concepts Addressed

Patterns: *Observed patterns in nature guide organization and classification and prompt questions about relationships and causes underlying them.*

- Patterns in rates of change and other numerical relationships can provide information about natural and human-designed systems.
- Patterns can be used to identify cause-and-effect relationships.

Cause and effect: *Events have causes, sometimes simple, sometimes multifaceted. Deciphering causal relationships, and the mechanisms by which they are mediated, is a major activity of science and engineering.*

- Cause-and-effect relationships may be used to predict phenomena in natural or designed systems.

Scale, proportion, and quantity: *In considering phenomena, it is critical to recognize what is relevant at different size, time, and energy scales, and to recognize proportional relationships between different quantities as scales change.*

- The observed function of natural and designed systems may change with scale.
- Phenomena that can be observed at one scale may not be observable at another scale.

Systems and system models: *A system is an organized group of related objects or components; models can be used for understanding and predicting the behavior of systems.*

- Systems may interact with other systems; they may have subsystems and be a part of larger complex systems.
- Models are limited in that they only represent certain aspects of the system under study.

> **CROSSCUTTING CONCEPTS**
>
> *A Framework for K–12 Science Education* describes seven crosscutting concepts as essential elements of a K–12 science and engineering curriculum. The learning progression for this dimension of the framework is addressed in volume 2, appendix G, of the *NGSS*. Elements of the learning progression for crosscutting concepts recommended for grades 6–8, as described in the performance expectations, appear after bullets below each concept.

HUMAN SYSTEMS INTERACTIONS — Framework and NGSS

Energy and matter: Tracking energy and matter flows into, out of, and within systems helps one understand their system's behavior.

- Within a natural (or designed system), the transfer of energy drives the motion and/or cycling of matter.

Structure and function: The way an object is shaped or structured determines many of its properties and functions.

- Complex and microscopic structures and systems can be visualized, modeled, and used to describe how their function depends on the shapes, composition, and relationships among its parts; therefore, complex natural and designed structures/systems can be analyzed to determine how they function.

Connections to the Nature of Science

- **Scientific knowledge is based on empirical evidence.** Scientific knowledge is based on logical and conceptual connections between evidence and explanations. Science disciplines share common rules of obtaining and evaluating empirical evidence.

- **Scientific knowledge is open to revision in light of new evidence.** The certainty and durability of scientific findings vary. Scientific findings are frequently revised and/or reinterpreted based on new evidence.

- **Science is a way of knowing.** Science is both a body of knowledge and processes and practices used to add to that body of knowledge. Scientific knowledge is cumulative and many people from many generations and nations have contributed to scientific knowledge. Science is a way of knowing used by many people, not just scientists.

- **Scientific knowledge assumes an order and consistency in natural systems.** Science assumes that objects and events in natural systems occur in consistent patterns that are understandable through measurement and observation. Science carefully considers and evaluates anomalies in data and evidence.

- **Science is a human endeavor.** Men and women from different social, cultural, and ethnic backgrounds work as scientists and engineers. Scientists and engineers rely on human qualities such as persistence, precision, reasoning, logic, imagination, and creativity. They are guided by habits of mind, such as intellectual honesty, tolerance of ambiguity, skepticism, and openness to new ideas. Advances in technology influence the progress of science, and science has influenced advances in technology.

CONNECTIONS
See volume 2, appendix H and appendix J, in the *NGSS* for more on these connections.

Connections to NGSS

- **Science addresses questions about the natural and material world.** Scientific knowledge is constrained by human capacity, technology, and materials. Science limits its explanations to systems that lend themselves to observation and empirical evidence. Scientific knowledge can describe consequences of actions but is not responsible for society's decisions.

Connections to Engineering, Technology, and Applications of Science

- **Interdependence of science, engineering, and technology.** Engineering advances have led to important discoveries in virtually every field of science, and scientific discoveries have led to the development of entire industries and engineered systems. Science and technology drive each other forward.

HUMAN SYSTEMS INTERACTIONS — Framework and NGS.

DISCIPLINARY CORE IDEAS

A Framework for K–12 Science Education has four core ideas in life sciences.

LS1: From molecules to organisms: Structures and processes

LS2: Ecosystems: Interactions, energy, and dynamics

LS3: Heredity: Inheritance and variation of traits

LS4: Biological evolution: Unity and diversity

The questions and descriptions of the core ideas in the text on these pages are taken from the NRC *Framework* for grades 6–8 to keep the core ideas in a rich and useful context.

The performance expectations related to each core idea are taken from the NGSS for middle school.

Disciplinary Core Ideas Addressed

The **Human Systems Interactions Course** connects with the NRC *Framework* 6–8 grade band and the NGSS performance expectations for the middle school grades. The course focuses on core ideas for life sciences primarily and physical science secondarily.

Life Sciences

Framework core idea LS1: From Molecules to Organisms: Structures and Processes—How do organisms live, grow, respond to their environment, and reproduce?

- **LS1.A: Structure and function**

  How do the structures of organisms enable life's functions? [All living things are made up of cells, which is the smallest unit that can be said to be alive. An organism may consist of one single cell (unicellular) or many different numbers and types of cells (multicellular). Unicellular organisms (microorganisms), like multicellular organisms, need food, water, a way to dispose of waste, and an environment in which they can live.

  Within cells, special structures are responsible for particular functions, and the cell membrane forms the boundary that controls what enters and leaves the cell. In multicellular organisms, the body is a system of multiple interacting subsystems. These subsystems are groups of cells that work together to form tissues or organs that are specialized for particular body functions.]

- **LS1.C: Organization for matter and energy flow in organisms**

  How do organisms obtain and use the matter and energy they need to live and grow? [Plants, algae (including phytoplankton), and many microorganisms use the energy from light to make sugars (food) from carbon dioxide from the atmosphere and water through the process of photosynthesis, which also releases oxygen. These sugars can be used immediately or stored for growth or later use. Animals obtain food from eating plants or eating other animals. Within individual organisms, food moves through a series of chemical reactions in which it is broken down and rearranged to form new molecules, to support growth, or to release energy. In most animals and plants, oxygen reacts with carbon-containing molecules (sugars) to provide energy and produce carbon dioxide; anaerobic bacteria achieve their energy needs in other chemical processes that do not require oxygen.]

Connections to NGSS

- **LS1.D: Information processing**
  How do organisms detect, process, and use information about the environment? [Each sense receptor responds to different inputs (electromagnetic, mechanical, chemical), transmitting them as signals that travel along nerve cells to the brain. The signals are then processed in the brain, resulting in immediate behaviors or memories. Changes in the structure and functioning of many millions of interconnected nerve cells allow combined inputs to be stored as memories for long periods of time.]

The following NGSS Grade 6–8 Performance Expectations for LS1 are derived from the Framework disciplinary core ideas above.

- MS-LS1-1. Conduct an investigation to provide evidence that living things are made of cells; either one cell or many different numbers and types of cells.
- MS-LS1-3. Use argument supported by evidence for how the body is a system of interacting subsystems composed of groups of cells.
- MS-LS1-7. Develop a model to describe how food is rearranged through chemical reactions forming new molecules that support growth and/or release energy as this matter moves through an organism.
- MS-LS1-8. Gather and synthesize information that sensory receptors respond to stimuli by sending messages to the brain for immediate behavior or storage as memories.

Physical Sciences

Framework core idea PS3: Energy—How is energy transferred and conserved?

- **PS3.D: Energy in chemical processes and everyday life**
  How do food and fuel provide energy? If energy is conserved, why do people say it is produced or used? [The chemical reaction by which plants produce complex food molecules (sugars) requires an energy input (i.e., from sunlight) to occur. In this reaction, carbon dioxide and water combine to form carbon-based organic molecules and release oxygen.

  Both the burning of fuel and cellular digestion in plants and animals involve chemical reactions with oxygen that release stored energy. In these processes, complex molecules containing carbon react with oxygen to produce carbon dioxide and other materials.]

  This core idea is secondary to MS-LS1-7 above.

Human Systems Interactions Course—FOSS Next Generation

HUMAN SYSTEMS INTERACTIONS — Framework and NGSS

FOSS NEXT GENERATION K–8 SCOPE AND SEQUENCE

Grade	Physical Science	Earth Science	Life Science
6–8	Waves* Gravity and Kinetic Energy* Chemical Interactions Electromagnetic Force* Variables and Design*	Planetary Science Earth History Weather and Water	Heredity and Adaptation* Human Systems Interactions* Populations and Ecosystems Diversity of Life
5	Mixtures and Solutions	Earth and Sun	Living Systems
4	Energy	Soils, Rocks, and Landforms	Environments
3	Motion and Matter	Water and Climate	Structures of Life
2	Solids and Liquids	Pebbles, Sand, and Silt	Insects and Plants
1	Sound and Light	Air and Weather	Plants and Animals
K	Materials and Motion	Trees and Weather	Animals Two by Two

* Half-length course

Full Option Science System

HUMAN SYSTEMS INTERACTIONS — *Materials*

Contents

Kit Inventory List 50

Materials Supplied by the Teacher 51

Important Information for First-Time FOSS Users 53

Preparing the Kit for Your Classroom 56

Care, Reuse, and Recycling 58

The Human Systems Interactions kit contains

- *Teacher Toolkit: Human Systems Interactions*

  1 *Investigations Guide: Human Systems Interactions*

  1 *Teacher Resources: Human Systems Interactions*

  1 *FOSS Science Resources: Human Systems Interactions*

- *FOSS Science Resources: Human Systems Interactions* (class set of student books)

- Equipment for 5 classes of 32 students

Each investigation in this course is divided into two to four parts. Each part has a Materials section that details the materials in the kit and the materials supplied by the teacher that will be used by each group of students and the class. The kit includes most of the learning equipment needed by students. There are enough consumable materials in the kit for 5 classes of 32 students each. Some of the teacher-supplied items can also be ordered through Delta Education.

For each investigation, you will need one computer with Internet access that can be displayed to the class, either by an LCD projector, interactive whiteboard, or large screen.

For updates to information on materials used in this course and access to the Safety Data Sheets (SDS), go to www.fossweb.com. Links to replacement-part lists and customer service are also available on FOSSweb.

▶ **NOTE**
Delta Education Customer Service can be reached at 1-800-258-1302.

Full Option Science System

HUMAN SYSTEMS INTERACTIONS — *Materials*

KIT INVENTORY List
Drawer 1—permanent equipment

Equipment condition

1	*Teacher Toolkit: Human Systems Interactions* (1 *Investigations Guide*, 1 *Teacher Resources*, and 1 *FOSS Science Resources: Human Systems Interactions*)	
32	*FOSS Science Resources: Human Systems Interactions*, student books ★	
1	Aluminum foil, roll, 23 m (25')	
16	Braille strips	
17	Card sets, Structural Levels, 8 cards/set	
50	Cups, plastic, 250 mL (9 oz.)	
1	Hole punch	
84	Labels, removable, 1 × 4.5 cm ✪	
36	Mirrors	
1	Poster, *Science Safety*	
1	Poster set, Systems, 8 posters/set	
4	Stopwatches	
1	String, ball, 30 m/ball	
1	Tape, painter's, roll	
3	Trays, cafeteria	
24	Vials, with caps, 12 dram	
5	Zip bags, 4 L ✪	

★ The student books are shipped separately in two boxes of 16 hardbound books each.

Drawer 1—consumable equipment

100	Cotton balls	
200	Index cards, 7.5 × 12.5 cm (3" × 5")	
100	Paper clips, jumbo	
500	Paper clips, regular	
500	Self-stick notes, 7.5 × 7.5 cm (3" × 3")	
8	Tape, transparent, rolls, 16.5 m/roll	

✪ These items might occasionally need replacement.

Full Option Science System

MATERIALS *Supplied by the Teacher*

Each part of each investigation has a Materials section that describes the materials required for that part. It lists materials needed for each student or group of students and for the class.

Be aware that you must supply some items. These are indicated in the materials list for each part of the investigation with an asterisk (★). Here is a summary list of those items. Some of the supplies and tools are available from Delta Education. Check the replacement-part list for the course on the Delta Education website.

Technology equipment
- Computers with Internet access
- 1 Document camera or overhead projector
- 1 Projection system
- Extension cords with multiple outlets (optional)

Measuring tools
- 32 Rulers (optional)

Paper
- Butcher paper, dark (optional)
- 12 Card stock, 22 × 28 cm (8.5" × 11") (optional)
- Chart paper
- Science notebooks (composition books)
- White paper, 28 × 44 cm (11" × 17")
- White paper, 22 × 28 cm (8.5" × 11")

Supplies
- Extracts/scents
- Glue sticks (optional)
- Nutrition labels
- Rubbing alcohol
- 8 Sheet protectors, plastic (optional)
- Transparent tape (optional)
- Water bottle with nutrition label

HUMAN SYSTEMS INTERACTIONS — *Materials*

Other tools

- 16 Calculators
- 32 Colored pencils, eight different colors, four of each color
- 1 Dollar bill, crisp
- 8 Erasers, whiteboard (optional)
- • Marking pens and highlighters
- 8 Marking pens, whiteboard (optional)
- 1 Memory-test set (key, paper bag, stapler, book, paper clip, ruler, comb, dollar bill, fork, cork)
- 8 Mini-whiteboards (optional)
- 32 Scissors

IMPORTANT *Information for First-Time FOSS Users*

If this is your first time using a FOSS middle school course, you should become familiar with a few items before beginning instruction. These steps will also prepare you to teach any other FOSS middle school course.

1. Plan for student notebooks

In FOSS, students keep science notebooks both as organized records of their scientific investigations and as places to reflect about their thinking. Notebook opportunities appear in each part of each investigation.

Students will need their own notebooks dedicated for use in science class, in which they can record focus questions, observations, data, conclusions, their own questions, and so on. These notebooks are typically bound composition books in which students make entries and glue or tape photocopied notebook sheets or other artifacts.

In preparation for each part of each investigation, you will make copies of the specified notebook masters. You can copy the preprinted notebook masters from *Teacher Resources* or download digital versions from www.FOSSweb.com. Each notebook master consists of two copies of a notebook sheet, so each photocopied page will need to be cut in half. Sometimes you might prefer to project a notebook master and have students copy some information from the notebook sheet into their notebooks, adding their own data and responses.

In the first investigation, make sure students have prepared their notebooks by setting up a table of contents, creating an index for vocabulary words, and numbering the pages. For more information on notebook use in FOSS, see the Science Notebooks in Middle School chapter.

> **TEACHING NOTE**
>
> *Notebook sheets are available on FOSSweb in several formats. For each notebook sheet, you can select "to photocopy," which will be identical to the printed notebook masters in **Teacher Resources**, or "to project," which is rotated and zoomed for easier display. You can also type into these notebook sheets while projecting them.*

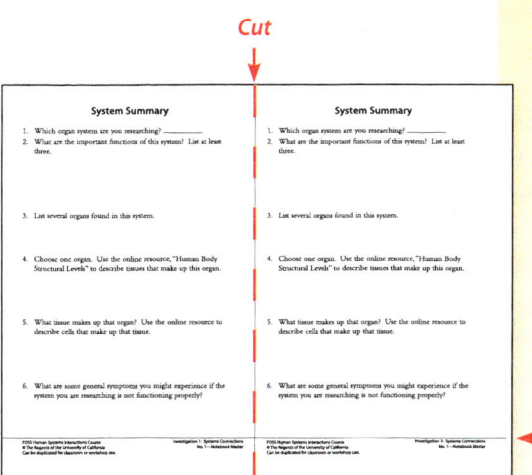

Notebook master

Human Systems Interactions Course, Second Edition

HUMAN SYSTEMS INTERACTIONS — *Materials*

2. Plan for online activities and projection

Throughout this course, you will need to project digital components through your computer for the class to see. The Getting Ready section for each part will indicate what to prepare.

In general, you will need regular access to a computer with Internet access, a document camera, and either an LCD projector or a large-screen display. If regular projection is difficult given your classroom setup, you could use the notebook masters and teacher masters to make transparencies for use with a document camera or an overhead projector.

For other projection needs, such as displaying a FOSSweb program, you will need to make sure students can see the computer display.

3. Become familiar with FOSSweb

If you have never logged into FOSSweb before, visit the site to set up your account. The site is used throughout the course to project teacher masters and notebook sheets, display digital components, such as animations and simulations, and provide student access to course resources and assignments that you create. For more information on how to set up an account and to access the digital resources, see the FOSSweb and Technology chapter in *Teacher Resources*.

Once you've logged in, familiarize yourself with the layout of the site and the additional resources available to you there. The easiest way to access resources is by clicking the icon for the course and going to "Resources by Investigation."

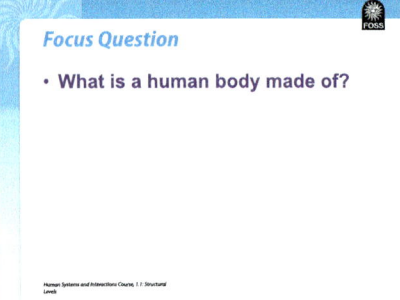

Teaching Slides

4. Review teaching slides

The teaching slides are a series of editable slides for you to use with your class as an instructional tool. There is one set of slides for each part of each investigation. Look for the teaching slides under Digital-Only Resources on FOSSweb.

5. Plan groups

Plan to organize students into groups of four around lab benches or tables. Seating should facilitate students' working together and sharing observations and ideas. The "for each group" section of the materials list will always describe the materials needed by a group of four students.

6. Display safety poster

Display the *Science Safety* poster in a prominent location in the classroom.

7. **Set up a materials station**

    Plan to establish a materials station where students will always pick up and return materials. Select a location that minimizes congestion and provides easy supervision as needed.

8. **Assess progress throughout the course**

    Embedded (formative) assessments provide a variety of ways to gather information about students' thinking while their ideas are developing. These assessments are designed to be diagnostic. They provide you with information about student learning so that you know if you need to plan a next step to clarify understanding before going on to the next part of the investigation. Each Getting Ready section describes an embedded-assessment strategy you may find useful in that part. Two assessment masters, *Embedded Assessment Notes* and *Performance Assessment Checklist*, are provided as tools to help you analyze students' data (see the Assessment chapter in *Teacher Resources* for more on how to use these tools). The *Performance Assessment Checklist* is in two formats, one for individual students and one for groups.

    At the end of most investigations, there is an I-Check benchmark assessment. The questions on these assessments are summative —they examine all the concepts students have learned up to that point in the curriculum. You can find out more about I-Check assessments in the Assessment chapter and in Investigation 2. Use the *Assessment Record* to record results. Check FOSSweb for downloadable spreadsheets for the *Perfomance Assessment Checklist* and *Assessment Record*.

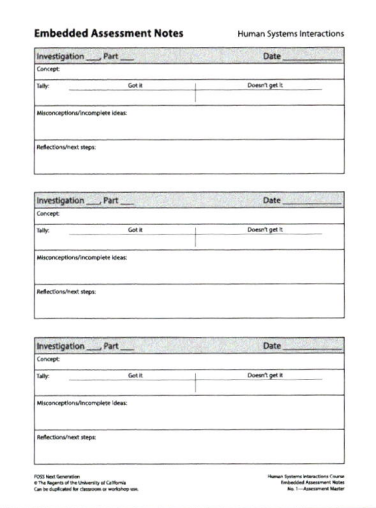

Embedded Assessment Notes

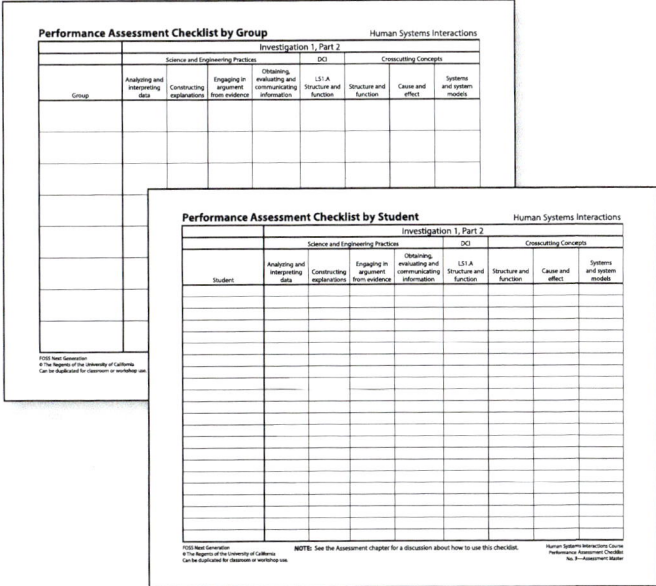

Performance Assessment Checklists

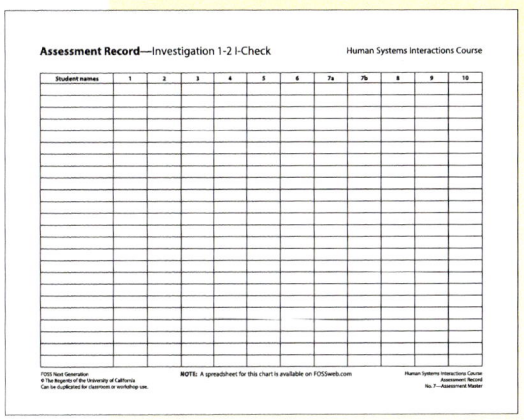

Assessment Record

Human Systems Interactions Course, Second Edition

55

HUMAN SYSTEMS INTERACTIONS — Materials

PREPARING the Kit for Your Classroom

Some preparation is required each time you use the kit. Doing things before beginning the course will make daily setup quicker and easier.

Each part of each investigation includes a section called Getting Ready, which describes what you need to do or consider to be prepared to conduct the part.

Note that a few items are consumable, but there should be enough in the kit for at least five classes before you need to restock.

One-Time Preparation

Some of the preparation will need to be done only once. Here are things that require one-time preparation.

Investigation 1, Part 1

Check Structural Levels card sets.

Investigation 2, Part 1

Gather eight food packages with nutrition labels (students can bring them in). You can use these from year to year. Obtain one plastic drinking bottle with a nutrition label.

Investigation 3, Part 2

Make a copy of teacher masters H–J, *Neural-Message Relay Cards A–C*, for each group of eight students (two groups of four will share one set). Prepare bags of group materials for the neural message relay.

Investigation 3, Part 3

Obtain eight different scents or extracts. Prepare scent vials using cotton balls and labels. The vials may need to be refreshed from year to year, using the same scents if at all possible.

▶ **SAFETY NOTE**
Be sure to find out if any students have allergies or sensitivities to extracts or scented oils.

Review Safety Guidelines

There is a safety poster in the kit. Consider how to introduce the class rules so that everyone has a safe science experience.

Reserve Computers

Students should have access to computers or tablets in pairs or groups throughout the course. This is especially important in these parts:

- Investigation 1, Part 1, for homework
- Investigation 1, Part 2, for group research
- Investigation 2, Part 1, for online activity
- Investigation 2, Part 2, for online activity
- Investigation 3, Part 1, for online activities
- Investigation 3, Part 3, for online activities

Plan ahead to use multiple computers at those times.

Sequential Classes

The materials are designed to be used with sequential classes. Organize a materials station in a central location in the classroom. Organize the materials at the station before first period. Each period, the appropriate materials are picked up for each group by a Getter, used for the investigation, inventoried by students at the end of the period, and returned to the materials station by a Getter. You can quickly review the materials station to ensure that all the materials came back (and take appropriate action if they didn't) and that the materials are ready for the next class.

HUMAN SYSTEMS INTERACTIONS — *Materials*

CARE, *Reuse, and Recycling*

When you finish teaching the course, inventory the kit carefully. Note the items that were used up, lost, or broken, and immediately arrange to replace the items. Use a photocopy of the *Kit Inventory List*, and put your marks in the "Equipment condition" column. Replacement parts are available for FOSS by calling Delta Education at 1-800-258-1302 or by using the online replacement-part catalog (www.DeltaEducation.com/foss/buy).

The items in the kit have been selected for their ease of use and durability. Make sure that items are clean and dry before putting them back in the kit. Small items should be inventoried (a good job for students under your supervision) and put into zip bags for storage. Any items that are no longer useful for science should be properly recycled.

INVESTIGATION 1 – Systems Connections

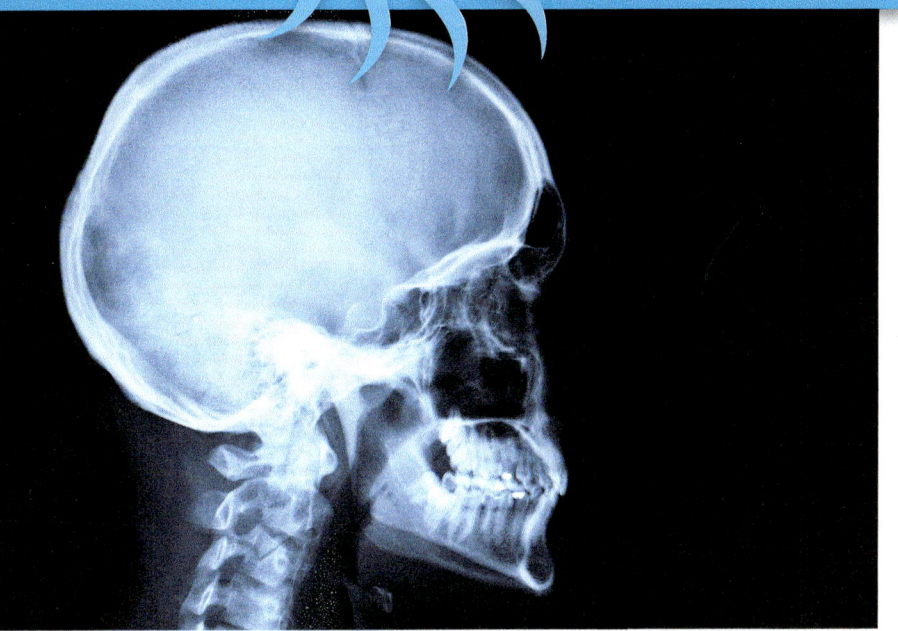

PURPOSE

Systems Connections prompts students to solve a mystery—a disease mystery. On the path to diagnosis, students encounter the levels of complexity in humans: cells form tissues, tissues form organs, organs form systems, and systems form a multicellular organism, the human. They discover how organ systems interact, each dependent on all the others for its functions.

Content

- Multicellular organisms are complex systems composed of organ systems, which are made of organs, which are made of tissues, which are made of cells.
- Cells are made of cell structures, which are made of molecules, which are made of atoms.
- The human body is a system of interacting subsystems (circulatory, digestive, endocrine, excretory, muscular, nervous, respiratory, skeletal, and others).

Practices

- Obtain, evaluate, and communicate information regarding a single human organ system.
- Diagnose a disease affecting a patient by evaluating research information and evidence.
- Engage in argument from evidence to defend conclusions.

Part 1
Human Body Structural Levels **72**

Part 2
Systems Research **83**

Science and Engineering Practices
- Asking questions
- Analyzing and interpreting data
- Constructing explanations
- Engaging in argument from evidence
- Obtaining, evaluating, and communicating information

Disciplinary Core Ideas
LS1: How do organisms live, grow, respond to their environment, and reproduce?
LS1.A: Structure and function

Crosscutting Concepts
- Cause and effect
- Scale, proportion, and quantity
- Systems and system models
- Structure and function

FOSS Full Option Science System

INVESTIGATION 1 – Systems Connections

Investigation Summary	Time	Focus Question and Practices
PART 1 **Human Body Structural Levels** Students are presented with a patient who has symptoms that could lead to a number of diagnoses. They determine a course of learning that begins with confirming the levels of complexity in a multicellular organism.	**Assessment** 1 Session **Active Inv.** 1 Session	**What is a human body made of?** **Practices** Asking questions Engaging in argument from evidence Obtaining, evaluating, and communicating information
PART 2 **Systems Research** Students continue their research to determine a diagnosis by focusing on human organ systems. They learn about how one organ system interacts with other organ systems in the body to support life processes. They pool their learning with the rest of the class. They conclude by making a tentative diagnosis of the patient, arguing their case to other students. They learn additional information that allows them to make a conclusive diagnosis.	**Active Inv.** 4 Sessions	**How do human organ systems interact?** **Practices** Analyzing and interpreting data Constructing explanations Engaging in argument from evidence Obtaining, evaluating, and communicating information

* A class session is 45–50 minutes.

At a Glance

Content Related to DCIs	Literacy/Technology	Assessment
• Multicellular organisms are complex systems composed of organ systems, which are made of organs, which are made of tissues, which are made of cells. • Cells are made of cell structures, which are made of molecules, which are made of atoms.	**Science Notebook Entry** Answer the focus question **Online Activities** "Structural Levels Cards" "Levels of Complexity" (optional) **Video** *Doctor Interview 1*	**Benchmark Assessment** *Entry-Level Survey* **Embedded Assessment** Science notebook entry
• Multicellular organisms are made of organ systems, which are made of organs, which are made of tissues, which are made of cells. • Cells are made of cell structures, which are made of molecules, which are made of atoms. • The human body is a system of interacting subsystems.	**Science Notebook Entry** *System Summary* *Systems Interactions* *Connect the Systems* **Science Resources Book** "Human Organ Systems" **Online Activities** "Structural Levels Cards" "Human Systems Structural Levels" **Video** *Doctor Interview 2*	**Embedded Assessment** Performance assessment **NGSS Performance Expectations addressed in this investigation** MS-LS1-1 (foundational) MS-LS1-3

Human Systems Interactions Course—FOSS Next Generation   61

INVESTIGATION 1 – Systems Connections

SCIENTIFIC and Historical Background

This background section appears in each investigation chapter. It is written for you. It is not for your students, nor should all the concepts discussed here be shared with them. The background discussion will reach deeper into the science, discussing ideas that may not be appropriate for your students. We will often use terminology that will not be introduced to students until much later. In some instances, connections to upcoming concepts will be discussed along with the misconceptions students may need to work through to make sense of it all. You need to have the big idea of where students are going, but they need to have these new ideas unfold before them in a logical progression so they can take it all in.

You will find teaching notes embedded in *Investigations Guide* that will point out where students may stumble or have questions—this is where the background can help you. Of course, middle school students have a knack for asking questions that you or the authors cannot anticipate. Hurray for them: questions begin a scientific quest.

© Monkey Business Images/Shutterstock

What Is a Human Body Made Of?

The human is a multicellular organism. This means that in order to carry out all of the functions of life such as eating food for energy, drinking water, exchanging gases such as oxygen and carbon dioxide, and excreting waste, human cells work together. The manner in which they accomplish this "working together" makes for a fascinating study.

We usually don't think of the myriad complex tasks our body takes care of every second of the day until something goes wrong. In this course, the study of disease opens the door for students into the study of the human body. Students are confronted with the **symptoms** of a mysterious disease and asked to help **diagnose** it. This motivates them to learn how the human body is organized and how it works.

"What is the human body made of?" Most middle school students will respond, "**Cells**." Scientists can distinguish between more than 200 different types of cells in the human body. But each of those unique types of cells are made of the same structures or components.

The human cell is eukaryotic: it has a nucleus which contains the genetic information (deoxyribonucleic acid, or DNA). It also has a cell membrane, which is a semipermeable barrier defining and isolating it from its environment, allowing some substances to move in and out and keeping other structures or substances safely inside or outside. The mitochondria transform the energy-rich **molecules** of the food we eat into

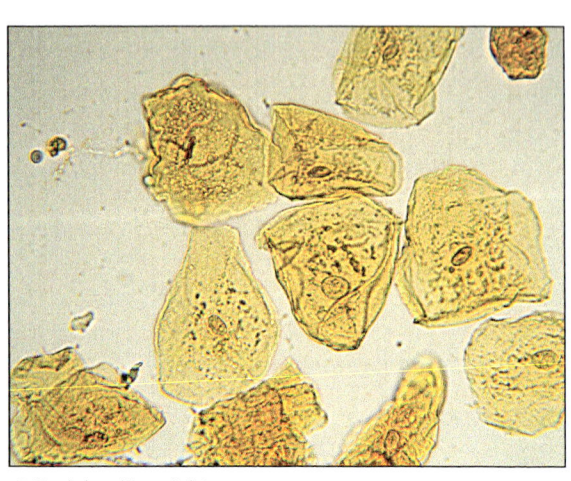
© iStockphoto/Nancy Nehring

62 Full Option Science System

Scientific and Historical Background

energy (molecules of adenosine triphosphate, or ATP) that the cell can use for growth and maintenance.

There are about 20 different types of organelles and **cell structures** in the cell, and every one is composed of complex molecules such as carbohydrates, proteins, and lipids (fats). Those molecules (made of **atoms**) come from the foods we eat.

Each human cell is descended from a single fertilized egg. How does one cell with one set of genetic instructions become trillions of different cells specialized for different functions in muscles, bones, the brain, and all the other organs? Since the decoding of the human genome in 2003, scientists have been able to determine the method that cells use to activate specific genes within the DNA. It turns out that regulatory chemical compounds and the proteins (histones) that enable the DNA to spool into its spiral form mark the DNA and thus turn genes on or off, controlling the production of proteins and leading to the development of different kinds of cells. These regulatory compounds and proteins are called the epigenome, and there is much exciting research into how it functions. This research is important in part because changes in the epigenome can cause diseases such as cancer.

Different cells have distinctly different functions to perform. Cells that perform the same function form **tissues**, such as connective tissue (including fat, blood, and bone), epithelial tissue (the tissue that lines internal and external body surfaces), muscle tissue, and nervous tissue.

Tissues form **organs** which perform a single function or several related functions. For instance, the heart is made of cardiac muscle tissue, the blood vessels within the heart are made of smooth muscle tissue, the valves are constructed of connective tissue, and nervous tissue in the heart regulates the heart rate. As different as all these tissues are, they work together to perform a single function—to pump blood throughout the body.

"How does one cell with one set of genetic instructions become trillions of different cells specialized for different functions in muscles, bones, the brain, and all the other organs?"

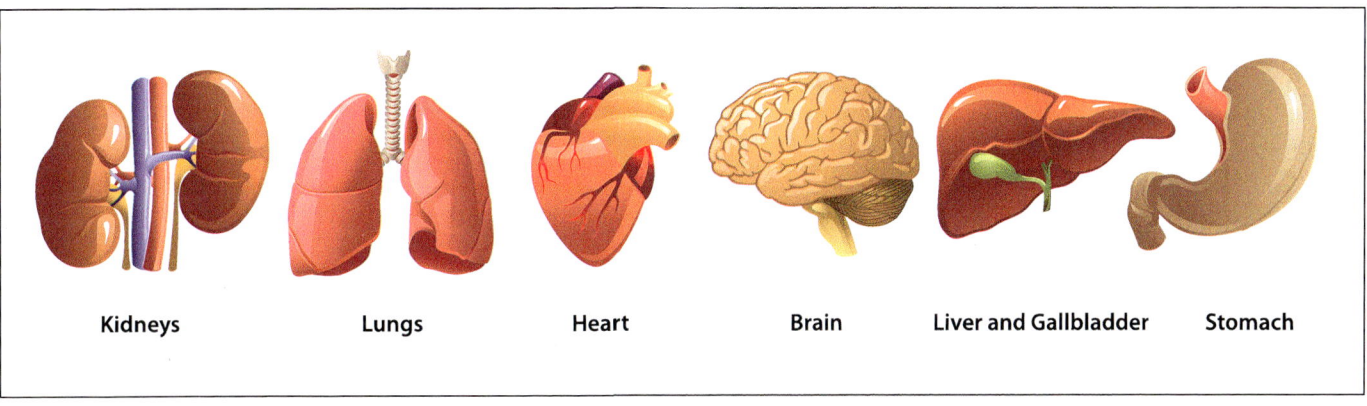

Kidneys   Lungs   Heart   Brain   Liver and Gallbladder   Stomach

© Sedova Elena/Shutterstock

INVESTIGATION 1 – Systems Connections

> **NOTE**
> The circulatory system can be descried as a larger system with two subsystems—the cardiovascular system and the lymphatic system. In this course, we focus on the cardiovascular subsystem of the circulatory system.

A group of organs working together to perform related functions is an **organ system**. The **circulatory system** comprises the heart, arteries, veins, and blood. This system transports oxygen, food molecules, and **hormones** to all the cells in the body; removes wastes such as carbon dioxide, urea, and dead cells; fights infection; and helps regulate body temperature.

Another example is the **skeletal system** as seen in the illustrations below.

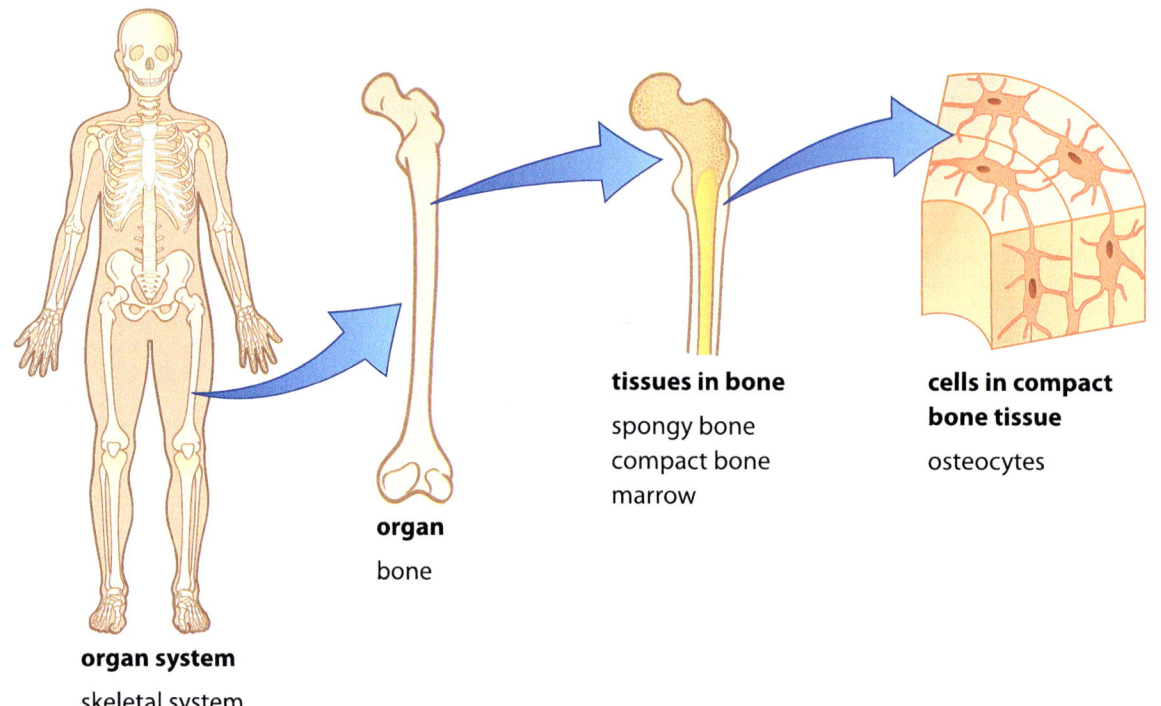

organ system
skeletal system

organ
bone

tissues in bone
spongy bone
compact bone
marrow

cells in compact bone tissue
osteocytes

How Do Human Organ Systems Interact?

The French physiologist Claude Bernard (1813–1878) said, "All the vital mechanisms, however varied they may be, have only one object, that of preserving constant the conditions of life in the internal environment." This implies correctly that every part of the complexity of the human body interacts to maintain what we call homeostasis.

In order for us to survive, all the organ systems must work together as a team. No system of the human body can work alone; they are all connected in some way. Students will research eight different systems: circulatory, digestive, endocrine, excretory, muscular, nervous, respiratory, and skeletal. (See the table for a summary.)

We will leave the reproductive, integumentary, lymphatic, and immune systems for high school.

Scientific and Historical Background

Circulatory System

Transports oxygen, nutrient molecules, and hormones to the cells in the body. Transports waste products such as carbon dioxide and water away from the cells.

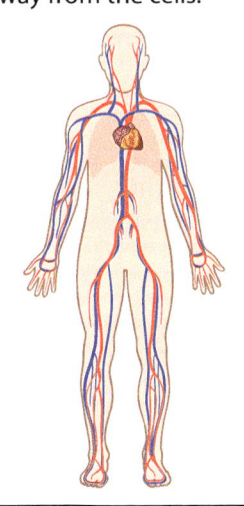

Digestive System

Breaks down food into nutrient molecules. Removes waste from the body.

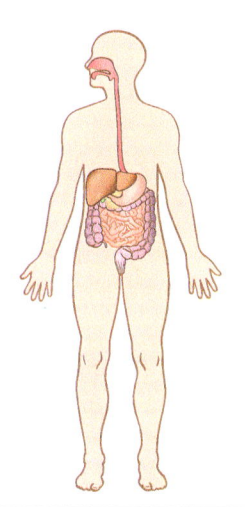

Endocrine System

Produces hormones, which travel through the body to regulate growth, metabolic, and sexual functions.

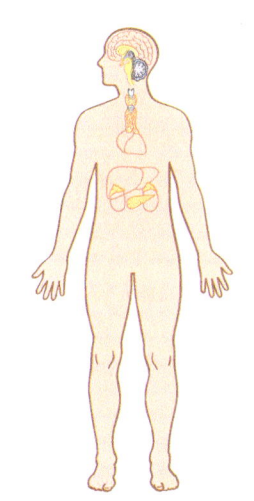

Excretory System

Removes waste products including excess water and salts from the body.

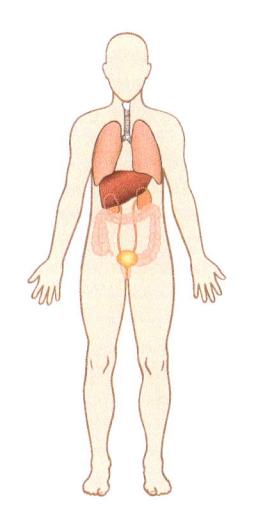

Muscular System

Skeletal muscles allow movement. Smooth muscles move substances through the body. Cardiac muscles in the heart pump blood.

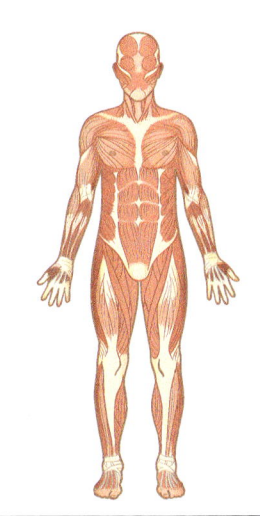

Nervous System

Controls all voluntary and involuntary actions in the body through the sending and receiving of electrical signals.

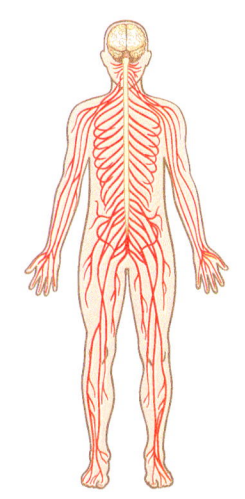

Respiratory System

Brings oxygen into the body and disposes of carbon dioxide and water.

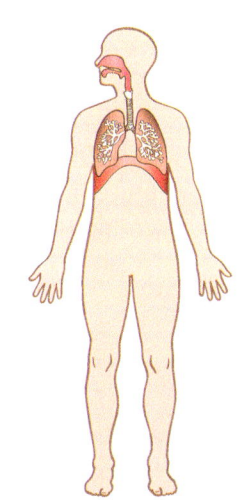

Skeletal System

Provides structure, shape, and protection. Allows movement and produces blood cells.

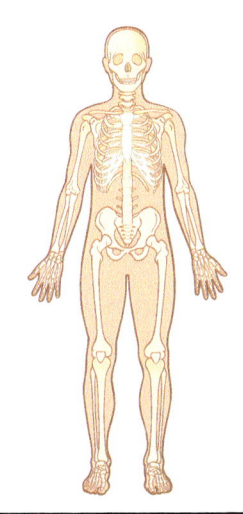

Human Systems Interactions Course—FOSS Next Generation

INVESTIGATION 1 – Systems Connections

Systems working together. Organ systems often work together to perform complicated tasks. Let's follow what happens when you eat a large meal. Contractions of the esophagus transport food to the stomach (both are muscular organs of the **digestive system**). The stomach muscles are stimulated by impulses from the enteric **nervous system**, which consists of millions of neurons (more neurons that the spinal cord!) embedded in the lining of the digestive system from the esophagus to the anus. As the stomach churns away, physically breaking down food into smaller pieces, the lining releases enzymes and acid to chemically break down the pieces into molecules. Hormones are also released, which help regulate the digestive process.

The digested food enters the small intestine for further digestion into food molecules. The food molecules cross through the intestinal wall into capillaries and the bloodstream. At this point, the digestive system enlists the aid of the nervous system. Impulses sent to the brain inform it of increased digestive activity. Blood flow increases to pick up food molecules for transport to the cells. The brain notifies the heart to pump more blood to the digestive system. The brain then responds, perceiving satiation and decreased interest in strenuous physical exercise. The blood that would go to the muscles is diverted to the digestive system. Mission accomplished.

The necessary communication to organize the complicated task of digestion is coordinated by the nervous and **endocrine systems**. But each organ system must do its part of the task. What happens when the work of an organ or system does not go according to plan?

Diabetes. An example of what happens to other systems when an organ does not function properly is **diabetes**. It is not well understood what causes type 1 diabetes, but it appears that susceptibility to the disease is inherited. Something triggers the immune system to attack insulin-producing beta cells in the pancreas. Insulin is an important hormone that helps move glucose into the body's tissues and cells. When insulin isn't produced, glucose cannot move into the cells and instead builds up in the bloodstream. High blood sugar and cell starvation are the result.

Kidneys ramp up their filtration efforts to dispose of the extra sugar, leading to increased urination, and potentially kidney failure. High blood sugar damages blood vessels and nerves. Increased blood pressure and the narrowing of arterial walls can lead to heart attack and stroke. Damage to capillaries that nourish the nerves can lead to nerve damage that initiates tingling, numbness, burning, and pain in the toes or fingers. This damage

> *"But each organ system must do its part of the task. What happens when the work of an organ or system does not go according to plan?"*

Scientific and Historical Background

can spread, eventually leading to loss of feeling in the affected limbs. Blindness due to damage of small blood vessels in the retina is another possible effect.

The consequences of the decrease of one hormone (insulin) in one organ of the body (the pancreas) are profound and multisystemic.

Disease is seldom limited to an effect on a single tissue or organ. Because the organ systems work together so intimately, each is vulnerable when another is affected. Knowing how the systems interact and how we affect our bodies with our lifestyle choices is important for our health.

Vocabulary
Atom
Cell
Cell structure
Circulatory system
Diabetes
Diagnosis
Digestive system
Endocrine system
Excretory system
Hormone
Molecule
Muscular system
Nervous system
Organ
Organ system
Respiratory system
Skeletal system
Symptom
Tissue

INVESTIGATION 1 – Systems Connections

TEACHING AND LEARNING about Systems Connections

Developing Disciplinary Core Ideas (DCI)

> **NGSS Foundation Box for DCI**
>
> **LS1.A: Structure and function**
> - All living things are made up of cells, which is the smallest unit that can be said to be alive. An organism may consist of one single cell (unicellular) or many different numbers and types of cells (multicellular).
> - In multicellular organisms, the body is a system of multiple interacting subsystems. These subsystems are groups of cells that work together to form tissues and organs that are specialized for particular body functions.

"My stomach hurts." "My friend is really sick." "I heard that my cousin started smoking." Like the rest of us, middle school students don't really think about their bodies until something is wrong. At this point in their physical development, however, their bodies are changing at a tremendous rate. Simultaneously, their social networks are expanding, morphing, and developing. So students are aware of not only themselves, but of their friends and families and the various interactions that make up their lives.

Students usually think about body function at a macroscopic level. Our heart beats faster when we exercise because our muscles need more blood. And we breathe deeper when we exercise because we get out of breath. This is the time to start students thinking about why those muscles need more blood and what getting out of breath really means. Muscles per se do not need more blood. It is the coordination of thousands of individual muscle cells that sound the alarm that calls for more resources. And getting out of breath is the body's response to a need for more ventilation, specifically the need to evacuate CO_2 at an increased rate. Everything in life from inspiration and exceptional physical performance to fatigue, disease, and dismay happens in cells. Life and death are measures of the success of cells and the systems that provide for their well-being.

One possible consequence of raising student's awareness of the cellular nature of their life is an emergent understanding of their responsibility for maintaining lifestyle behaviors that ensure high-quality performance of the multiple systems that ensure the healthy functioning of their cells. Students are at an age where they can start to make lifestyle choices independently. One of the poorest health decisions a person can make is to adopt the habit of smoking tobacco. A compromised respiratory system can lead to chronic reduction of the smooth flow of oxygen to the cells. Heart health is obviously essential to the continuation of healthy cells; it is the heart that drives the delivery of nutrients to and wastes from the cells. A heart compromised by excessive body weight is antithetical to healthy cells. These are potential topics for discussion, and should not be considered out of bounds. They do, however, require particular sensitivity.

The experiences students have in this investigation contribute to the disciplinary core idea **LS1.A: Structure and function.** In addition to teaching students about the core idea about the levels of complexity of the human body, each investigation in the **Human Systems Interactions Course** provides students with opportunities to experience two other dimensions of the scientific enterprise—the dimension of science and engineering practices and the dimension of crosscutting concepts.

Teaching and Learning about Systems Connections

Engaging in Science and Engineering Practices (SEP)

Engaging in the practices of science helps students understand how scientific knowledge develops; such direct involvement gives them an appreciation of the wide range of approaches that are used to investigate, model, and explain the world. Engaging in the practices of engineering likewise helps students understand the work of engineers, as well as the links between engineering and science. Participation in these practices also helps students form an understanding of the crosscutting concepts and disciplinary ideas of science and engineering; moreover, it makes students' knowledge more meaningful and embeds it more deeply into their worldview. (National Research Council, *A Framework for K–12 Science Education*, 2012, page 42)

The focus questions and notebook sheets provide scaffolds for students in this investigation that can be carefully removed in later investigations, as students become more experienced engaging in practices. In this first investigation, students engage in these practices.

- **Asking questions** that will provide clarifying information to support a diagnosis of a medical condition.
- **Analyzing and interpreting data** provided by a patient concerning health-related symptoms that will provide evidence for a diagnosis.
- **Constructing explanations** by applying scientific ideas to determine the relationship between structural levels in multicellular organisms; apply evidence to explain the cause of a medical condition.
- **Engaging in argument from evidence** to support an explanation for a medical diagnosis.
- **Obtaining, evaluating, and communicating information** by integrating technical information from text with that in media and visual displays to understand how human organ systems interact and what can go wrong; communicate understanding of human organ systems.

NGSS Foundation Box for SEP

- **Ask questions** to identify and/or clarify evidence and/or the premise(s) of an argument.
- **Analyze and interpret data** to provide evidence for phenomena.
- **Apply scientific ideas, principles, and/or evidence** to construct, revise, and/or use an explanation for real-world phenomena, examples, or events.
- **Respectfully provide and receive critiques** about one's explanations, procedures, models, and questions by citing relevant evidence and posing and responding to questions that elicit pertinent elaboration and detail.
- **Construct, use, and/or present an oral and written argument** supported by empirical evidence and scientific reasoning to support or refute an explanation or a model for a phenomenon (or a solution to a problem).
- **Integrate qualitative and/or quantitative scientific and/or technical information** in written text with that contained in media and visual displays to clarify claims and findings.
- **Communicate scientific and/or technical information** (e.g., about a proposed object, tool, process, system) in writing and/or through oral presentations.

INVESTIGATION 1 – Systems Connections

NGSS Foundation Box for CC

- **Cause and effect:** Cause-and-effect relationships may be used to predict phenomena in natural or designed systems. Phenomena may have more than one cause, and some cause-and-effect relationships in systems can only be described using probability.
- **Scale, proportion, and quantity:** The observed function of natural and designed systems may change with scale. Phenomena that can be observed at one scale may not be observable at another scale.
- **Systems and system models:** Systems may interact with other systems; they may have subsystems and be a part of larger complex systems.
- **Structure and function:** Complex and microscopic structures and systems can be visualized, modeled, and used to describe how their function depends on the shapes, composition, and relationships among its parts; therefore, complex natural and designed structures/systems can be analyzed to determine how they function.

Exposing Crosscutting Concepts (CC)

The third dimension of instruction involves the crosscutting concepts, sometimes referred to as the unifying principles, themes, or big ideas, that are fundamental to the understanding of science and engineering.

These concepts should become common and familiar touchstones across the disciplines and grade levels. Explicit reference to the concepts, as well as their emergence in multiple disciplinary contexts, can help students develop a cumulative, coherent, and usable understanding of science and engineering. (National Research Council, 2012, page 83)

In this first investigation, the focus is on these crosscutting concepts.

- **Cause and effect.** A good medical diagnosis requires understanding the relationship between the effect (symptoms) and its cause.
- **Scale, proportion, and quantity.** Multicellular organisms can be observed and described at a number of levels or scales, from cell organelles to cells, to tissues, to organs, to organ systems.
- **Systems and system models.** Humans are a collection of interacting, coordinated organ subsystems.
- **Structure and function.** The function of organ system structures (cells, tissues, organs) depends on the relationships among the parts.

Connections to the Nature of Science

- **Scientific knowledge is based on empirical evidence.** Scientific knowledge is based on logical and conceptual connections between evidence and explanations. Science disciplines share common rules of obtaining and evaluating empirical evidence.
- **Scientific knowledge is open to revision in light of new evidence.** The certainty and durability of scientific findings vary. Scientific findings are frequently revised and/or reinterpreted based on new evidence.
- **Scientific knowledge assumes an order and consistency in natural systems.** Science assumes that objects and events in natural systems occur in consistent patterns that are understandable through measurement and observation. Science carefully considers and evaluates anomalies in data and evidence.
- **Science is a human endeavor.** Scientists and engineers rely on human qualities such as persistence, precision, reasoning, logic, imagination, and creativity. They are guided by habits of mind, such as intellectual honesty, tolerance of ambiguity, skepticism, and openness to new ideas.

Teaching and Learning about Systems Connections

Conceptual Flow

The conceptual flow starts in Part 1 with the introduction of a patient describing symptoms of a medical condition to a doctor. Students generate a list of questions for the patient and then focus on the structural levels of the human body to understand how to approach the diagnosis of the disease. Students confirm that the **basic unit of life is the cell** and that all organisms are made of **one or more cells**. Students sort and organize a set of structural level cards by scale: **Multicellular organisms** are complex structures composed of **organ systems**, which are **made of organs**, which are **made of tissues** which are **made of cells**. Cells are **made of cell structures** (organelles), which are **made of molecules**, which are **made of atoms** (smallest scale).

In Part 2, students become specialists in one of the human organ systems—**circulatory, digestive, endocrine, excretory, muscular, nervous, respiratory,** and **skeletal**. (There are a few organ systems that are not included in this study that can be studied as extensions—immune, integumentary, lymphatic, and reproductive.) The class works together to describe the many ways that the **human organ systems interact** and the ways that a **disease** or condition (such as **diabetes**) can impact several systems. The part ends with the students using the information they have gathered to diagnosis the patient's disease—**hantavirus**.

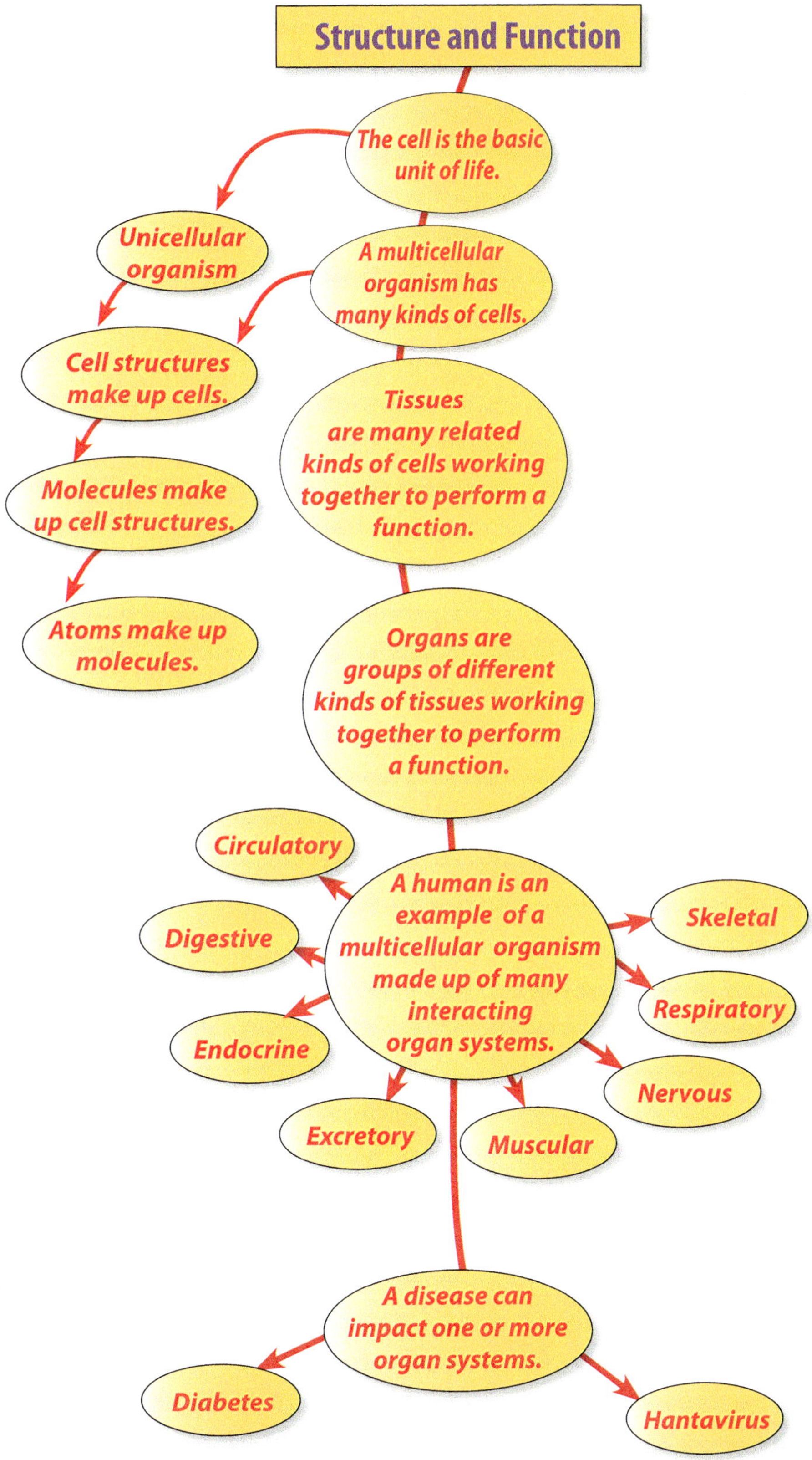

Human Systems Interactions Course—FOSS Next Generation

71

INVESTIGATION 1 – Systems Connections

MATERIALS for
Part 1: Human Body Structural Levels

Provided equipment
For each group
- 2 Structural Levels card sets

For the class
- 1 *Science Safety* poster

Teacher-supplied items
For each student
- 1 Science notebook (composition book)

For the class
- Chart paper
- 1 Marking pen
- 1 Document camera (optional)

FOSSweb resources
For each student
- *Entry-Level Survey*

For the class
- Video, *Doctor Interview 1*
- Online resource: "Structural Levels Cards"
- Online activity: "Levels of Complexity"

For the teacher
- 1 *Embedded Assessment Notes*
- Teaching slides, 1.1

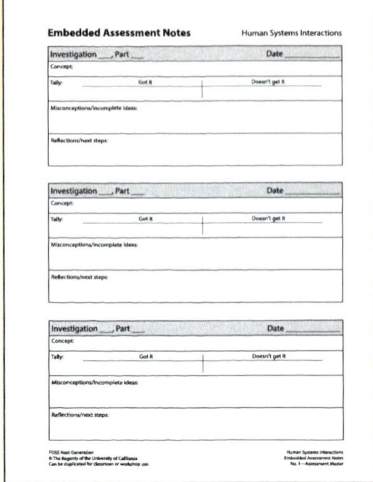

Embedded Assessment Notes

Part 1: Human Body Structural Levels

GETTING READY for
Part 1: Human Body Structural Levels

Quick Start

Schedule	1 session *Entry-Level Survey* 1 session active investigation
Preview	• Preview the FOSSweb Resources by Investigation for this part (such as printable masters, teaching slides, and multimedia) • Preview the video: *Doctor Interview 1*, Step 3 (*Doctor Interview 2* is shown in Part 2) • Preview online resource: "Structural Levels Cards," Step 7 • Preview/plan for homework: online activity, "Levels of Complexity," Step 13
Print or Copy	**For each student** • *Entry-Level Survey* **For the teacher** • *Embedded Assessment Notes*
Prepare Material	• Prepare for initial use and assessment **A** • Plan for student notebooks **B** • Preview Structural Levels card sets **C**
Plan for Assessment	• Review Step 12, "What to Look For" in the notebook entry

▶ **NOTE**
Schedule access to computers for Part 2.

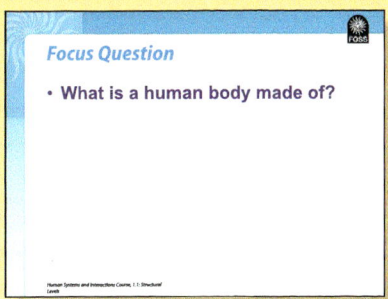

▶ **NOTE**
Preview the teaching slides on FOSSweb for this part.

Human Systems Interactions Course—FOSS Next Generation

73

INVESTIGATION 1 – Systems Connections

Preparation Details

A▶ Prepare for initial use and assessment

If this is your first use of the course, review the Materials chapter for critical information about accessing online resources, organizing the classroom and materials stations, and planning for materials. Review the Assessment chapter to plan for embedded, performance assessments, and benchmark assessments.

If you have never logged into FOSSweb before, visit the site to set up your account. For more information on how to set up an account and to access the digital resources, see the FOSSweb and Technology chapter in *Teacher Resources*. Once you've logged in, check the course page for any course updates. Preview any resources for this part by going to "Resources by Investigation."

B▶ Plan for student notebooks

In preparation for each part of each investigation, you will make copies of the specified notebook masters. In this first investigation, make sure students have prepared their notebooks by setting up a table of contents, creating an index for vocabulary words, and numbering the pages. For more information on notebook use in FOSS, see the Science Notebooks in Middle School chapter.

Students encounter new vocabulary in this part. Encourage students to define the new vocabulary in their own words, use labeled drawings to help them make sense of the terms, and add the terms to the indexes in their notebooks.

C▶ Preview Structural Levels card sets

Each pair of students will need a set of Structural Levels cards in Step 6 of Guiding the Investigation. A complete set includes six different cards packaged in a zip bag when the kit is new. Check the sets to make sure all are complete. You can find replacement cards by going to "Resources by Investigation" on FOSSweb.

TEACHING NOTE

Notebook entries serve as assessment opportunities for learning. Research shows that if you spend time reviewing student work each day and use that information to guide next instructional steps, students achieve significantly more. Use the Reflective Assessment Practice to make this a quick and easy process. Choose a random sample across your classes and plan 15 minutes to review student work. The important thing is that you are looking at student work on a daily basis as much as possible.

Part 1: Human Body Structural Levels

GUIDING *the Investigation*
Part 1: *Human Body Structural Levels*

SESSION 1

Students will . . .
- Express prior knowledge on *Entry-Level Survey* (Step 1)
- Obtain medical information about a sick patient from a video (Steps 2, 3)
- Generate a list of questions to gather more information (Steps 4, 5)
- Sort and organize a set of human body structural level cards (Steps 6–9)
- Review vocabulary and answer the focus question (Steps 10–12)
- Engage with online activity "Levels of Complexity" as homework (Step 13)

FOCUS QUESTION
What is a human body made of?

SESSION 1   45–50 minutes

1. Assess prior knowledge: *Entry-Level Survey*

Survey the students' understanding before instruction using the *Entry-Level Survey* assessment. The *Entry-Level Survey* is administered before instruction begins. It can be used to help you determine what students already know and what they need to learn, so that you can better plan how to focus each lesson. Encourage students to answer the *Entry-Level Survey* questions as best they can. Even if they think they don't know the answers, they should try to think about something related that they do know and apply that knowledge.

Collect the completed *Entry-Level Surveys*, look at them for diagnostic purposes, but don't make any marks or grade them. You can give them back to students now or at the end of the course so students can reflect on how their thinking changes over time.

2. Introduce disease *symptoms*

Ask students to think of a time when they were sick. What were the **symptoms** (signs that they were sick)? Who made a **diagnosis** (the identification of the disease)? After a moment, have students take turns sharing their experiences with a partner. Call on a few volunteers to share what they or their partner shared.

Tell students to imagine that you were approached by a friend who is a doctor. She has a patient who has some complicated symptoms, and she wants to take the case to "grand rounds." Grand rounds are when a doctor presents a patient's medical symptoms to other doctors and medical students to share experiences and to ask for assistance in finding a diagnosis for the patient's illness.

EL NOTE

Start a word wall for this investigation and add the words in bold.

EL notes present an opportunity for you to differentiate instruction for your English language learners and other students who may need additional language supports. You may decide to implement strategies depending upon your students' needs.

Human Systems Interactions Course—FOSS Next Generation   75

INVESTIGATION 1 – Systems Connections

ELA CONNECTION

This suggested strategy addresses the Common Core State Standards for ELA for literacy.

WHST 8: Gather relevant information from multiple print and digital sources.

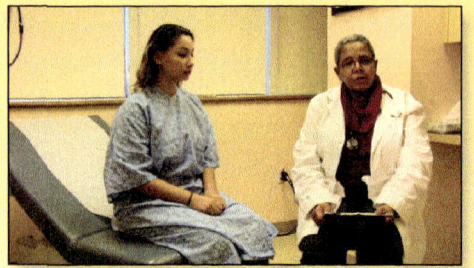

EL NOTE

Point out that the root word for "pulmonary" comes from the Latin "pulmon," which Spanish-speakers will recognize as the word for "lung."

ELA CONNECTION

This suggested strategy addresses the Common Core State Standards for ELA for literacy.

L8.4.b: Use Greek or Latin affixes and roots as clues to the meaning of a word.

SCIENCE AND ENGINEERING PRACTICES

Asking questions

Obtaining, evaluating, and communicating information

Tell students that you have a short video of the doctor's concerns. Ask them to open their notebooks to a clean page and write "Patient Symptoms and Possible Diseases" at the top. If students have not yet set up their notebooks, have them set up a table of contents and an index (for vocabulary words), and number the pages.

Tell student that they will play the role of medical students; therefore they will need to jot down the symptoms they hear and the diseases the doctor is considering.

3. Introduce the mystery disease

Project the online video *Doctor Interview 1*. If necessary, play it a second time. Give students a few minutes to compare their notes with a partner. Ask,

➤ *What are the patient's symptoms?* [Fever, fatigue (tiredness), muscle aches (especially the back), joint pain, and severe headaches.]

Write the symptoms on a piece of chart paper or on the board. Confirm that all students have written them down.

Ask,

➤ *What are the diseases that the doctor suspects?* [Lupus, hantavirus pulmonary syndrome, and Lyme disease.]

Tell students that they will learn more about the suspected diseases in a few sessions, but for now, they should know that *pulmonary* means having to do with the lungs and *syndrome* refers to a set of symptoms caused by one disease.

4. Generate a list of questions

Ask students to consider how they might go about working with the doctor. Tell students to work with their groups to come up with a list of questions to ask both the doctor and the patient. Have a Recorder in each group record his/her group's questions.

After a few minutes, call on the Recorders and record the questions on a sheet of chart paper to return to throughout the investigation. Use one sheet for each class.

Review the task and class list of questions.

➤ *What questions should you ask the doctor or patient?* [Where the patient has been, if anyone else has the disease, more about the human body, more about the diseases themselves such as causes and vectors, and so on.]

Part 1: Human Body Structural Levels

Suggest that the place to start is to understand more about the human body so that students can apply their knowledge to the mystery. In order to do that, they'll first determine what the body is made of.

5. **Focus question: What is a human body made of?**
Introduce the focus question. Write or project it on the board. Ask students to leave a page for additional information and notes on the mystery disease and turn to a clean page. They should write the focus question at the top.

➤ *What is a human body made of?*

Give students a minute to discuss the question in their groups and jot down their first thoughts in their notebooks. Call on a few groups to share what they think, and list the ideas on the front board. They will most likely share things such as obvious organs and body parts.

Students should leave space in their notebooks so they can return to the focus question at the end of their investigation into structural levels.

E L   N O T E

For students who need scaffolding, have them draw a picture of the human body and draw and/or label the parts they know.

6. **Introduce the Structural Levels cards**
Ask students to look at what the class has listed. If cells have not been a part of the discussion, ask,

➤ *What are all organisms made of?* [**Cells**.]

Once it has been established that all organisms, including humans, are made of cells, introduce the Structural Levels cards. Hold up or project the cell card and explain that cells are made of even smaller structures and that cells make up other structures in multicellular organisms (organisms that are made of more than one cell).

Say,

In order to understand how disease affects the human body, we need to consider all structural levels of the body and understand how they are related. This should help us understand how symptoms relate to structures in the body.

These cards will help us think about the relationship between different subsystems in the human body.

E L   N O T E

Add the word "cell" to the word wall.

At the beginning of a course, you can model how to make notebook entries by keeping a class notebook and displaying it using a document camera.

TEACHING NOTE

This content discussion should be a review for most students. Observe your students' facility with the academic vocabulary and concepts and focus the discussion accordingly.

Human Systems Interactions Course—FOSS Next Generation

INVESTIGATION 1 – Systems Connections

Describe the task. Working with a partner, students should take the cards, read the information on each, and lay them out in order of how they are related. Suggest that they put the most fundamental (or smallest) structure on the left and place the other cards in order, moving to the right. Once they are done, they should record their order on a new page in their notebooks. They need only write the name of the card, not all of the information.

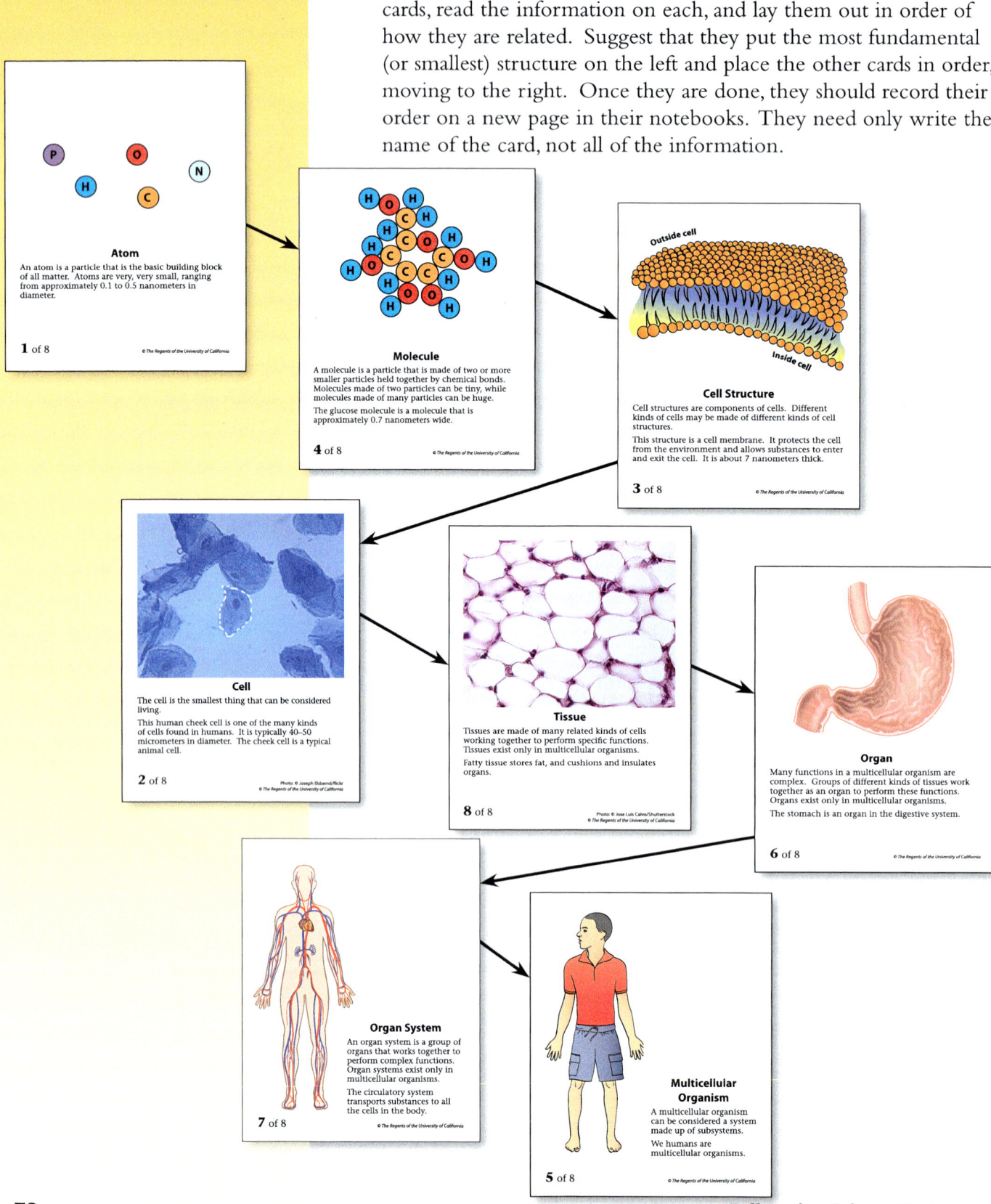

Part 1: Human Body Structural Levels

7. Compare within groups

Once pairs have had a few minutes to put their cards in order, have them compare their sequences with the other pair in their group. They should discuss their reasoning and make any changes they feel necessary (they might be happy with their own order). Ask one pair to share their sequence and either tape cards on the board or write the names in the order given. Students can also use the online resource "Structural Levels Cards" to display to the class the sequence of their cards. Ask students if they agree or disagree with the order. Have a volunteer explain why they agree. If a student disagrees, ask him/her to explain their reasoning.

If the correct order is quickly arrived at, confirm it and move directly to Step 8. Have students write the correct order in their notebooks.

If there is much discussion about the correct order, ask students to think about the crosscutting concept of scale and explain that this is an example of how we can study the human body at different levels of scale or complexity. Use these questions to focus on scale.

➤ *Which is the smallest unit of matter?* [Atom.]

➤ *Which card represents the largest unit of matter?* [Human.]

➤ *Which cards are you sure are positioned in proper relationship to each other?*

SCIENCE AND ENGINEERING PRACTICES
Engaging in argument from evidence

CROSSCUTTING CONCEPTS
Scale, proportion, and quantity

8. Confirm structural levels

Tell students to record information for each of the structures in their notebook during the discussion. (**NOTE:** If students have this information in their notebooks from the **FOSS Diversity of Life Course**, have them locate those entries, review, and add new information.)

➤ *Find the atom card. Figure out with your partner what the letters on the circles represent.* [Phosphorus, hydrogen, oxygen, carbon, and nitrogen.]

Point to each letter and have students call out the name as you write the atoms on the board. Tell students that these are the most common atoms (or elements) in biological substances.

E L   N O T E

Hold up or project the cards one by one and write down the information using a document camera or chart paper.

➤ *Atoms combine to make _____.* [**Molecules**.]

➤ *Find the molecule card. The molecule on the card is glucose. Name the atoms that make up glucose.* [Carbon, oxygen, and hydrogen.]

The most important biological molecules include carbohydrates, fats, proteins, and nucleic acids.

Ask students to hold up the card that molecules combine to make cell structures.

Human Systems Interactions Course—FOSS Next Generation

INVESTIGATION 1 – Systems Connections

TEACHING NOTE

Students may know the term "organelle" to mean the structures that make up cells. If they do, use that term.

➤ **Cell structures** are parts of cells. What is the cell structure on the card? [The cell membrane.]

➤ Discuss with your partner other cell structures you recall. [Responses should include nucleus, mitochondria, ribosomes, and so on.]

➤ Let's go to the next level. What do cell structures build? [Cells.]

➤ What do you know about cells? [Cells make up all life. Some life is single-celled, some organisms are made of trillions of cells. There are many kinds of cells.]

➤ What do related cells make up? [Tissues.] What do **tissues** do? [They perform specific functions. For example, fatty tissue stores fat.]

➤ Why don't single-celled organisms have tissues? [Single-celled organisms have only one cell.]

Tell students that there are four main types of tissue in humans (muscle, epithelial, nervous, and connective).

➤ Hold up the card that represents the kind of structures that are made of tissues. [Organs.] The stomach is one **organ**; what others can you think of?

➤ What are made of organs? [Organ systems.] What is the **organ system** on the card? [Circulatory system.]

➤ What other organ systems do you know of in the human body? [It is not important for students to make a long list here as they will be doing that in Part 2. Do not spend too much time on this.]

Tell students that the human body is itself a system that is made up of many subsystems: all the organ systems.

TEACHING NOTE

The human body's circulatory system is actually made up of two systems, the cardiovascular system (heart, blood vessels, and blood) and the lymphatic system which transports lymph throughout the body. In this course, we study the part of the circulatory system that includes the cardiovascular subsystem only. Students can find out about the lymphatic system through the extensions.

9. **Clean up**

Ask each pair of students to mix up the cards and return them in random order to the zip bags. Have Getters return the bags to the materials station.

10. **Review vocabulary**

Take a few minutes to review the vocabulary developed in this part. Ask students to highlight the vocabulary words and definitions in their notebooks and to update their vocabulary indexes if they haven't already done so. More information about vocabulary in notebooks is in the Science Notebooks in Middle School chapter.

atom
cell
cell structure
diagnosis
molecule
organ
organ system
symptom
tissue

80

Full Option Science System

Part 1: Human Body Structural Levels

11. Answer the focus question

Have students return to the page where they recorded the focus question.

➤ *What is the human body made of?*

Ask students to draw a line of learning under their initial ideas and answer the focus question using what they learned about the levels of scale and organization during the session working with the Structural Levels cards.

12. Assess progress: notebook entry

After students complete their responses to the focus question, collect a sample of notebooks from each class, and assess students' progress. The sample you select should give you a snapshot of the range of student understanding at this point in time. Ask students to turn in their notebooks open to the page you will be looking at.

What to Look For

- *The human body can be described at different scales or levels of organization that reflect increasing complexity.*
- *Students list the levels in order of scale and complexity: atoms, molecules, cell structures, cells, tissues, organs, organ systems.*

Plan to spend 15 minutes reviewing the selected sample of student responses. Using *Embedded Assessment Notes* as a tool, review the responses, and tally the number of students who got it or didn't get the concept. Record any alternative concepts that are evident, and decide if a next-step strategy is required before moving forward.

13. Extend the investigation with homework

For homework, have students access the "Levels of Complexity" online activity on FOSSweb. Ask them to explore the activity and compare the tissues, organs, and organ systems of the human and the insect. Ask students to gather information to answer this question.

➤ *How are the levels of the human and insect alike and how are they different?*

Suggest students use a comparison graphic organizer such as a box-and-T chart (see *Teacher Resources*, Science Notebooks in Middle School, Making Sense of Data, Graphic Organizers) to answer the question. They can then use the information to write a short paragraph explaining the comparisons.

TEACHING NOTE

See the Science Notebooks in Middle School chapter for additional information about using a line of learning.

TEACHING NOTE

More information about using Embedded Assessment Notes and examples of next-step strategies are in the Assessment chapter.

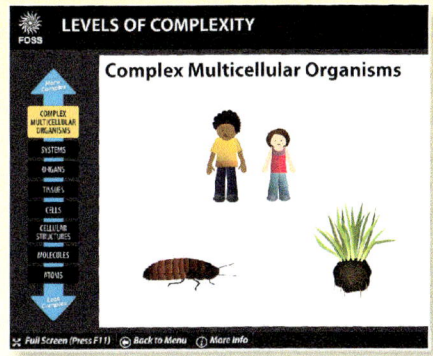

Human Systems Interactions Course—FOSS Next Generation

81

INVESTIGATION 1 – Systems Connections

ELA CONNECTION

These suggested strategies address the Common Core State Standards for ELA for literacy.

WHST.8: Gather relevant information from multiple print and digital sources.

L 4: Determine or clarify meaning of unknown words or phrases.

WRAP-UP/WARM-UP

14. **Review notebook entries**

    Give students a few minutes to review their notebook entries and ask questions to clarify the meaning of the vocabulary words introduced in this part. Encourage students to elaborate on their definitions and to add illustrations. Later, they can add these words to their index pages.

Part 2: Systems Research

MATERIALS for
Part 2: Systems Research

Provided equipment

For each student
- 1 *FOSS Science Resources: Human Systems Interactions*
- • "Human Organs Systems"
- • "Disease Information" from Images and Data section
- • "Diabetes Affects Human Organ Systems" from Images and Data section

For the class
- 1 Organ System card (from Structural Levels cards)
- 8 Systems posters
- • Self-stick notes, large, 7.5 × 7.5 cm
- • Painter's tape

Teacher-supplied items

For each group
- 2 Computers or tablets with Internet access
- • Colored pencils (four of eight different colors)
- 1 Mini-whiteboard, marking pen, eraser
- • Self-stick notes (for marking reading)
- 4 Rulers (optional)
- 1 Piece of card stock, 22 × 28 cm (optional)
- 1 Plastic sheet protector (optional)

For the class
- • Chart paper

FOSSweb resources

For each student
- 1 Notebook sheet 1, *System Summary*
- 1 Notebook sheet 2, *Systems Interactions*
- 1 Notebook sheet 3, *Connect the Systems*

For the class
- • Online resource: "Structural Levels Cards"
- • Online activity: "Human Systems Structural Levels"
- • Video: *Doctor Interview 2*

For the teacher
- • *Performance Assessment Checklist*
- • Teaching slides, 1.2

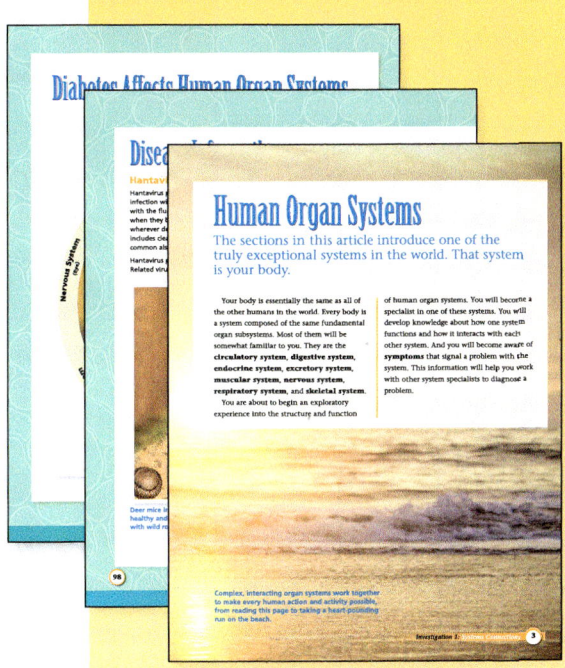

FOSS Science Resources

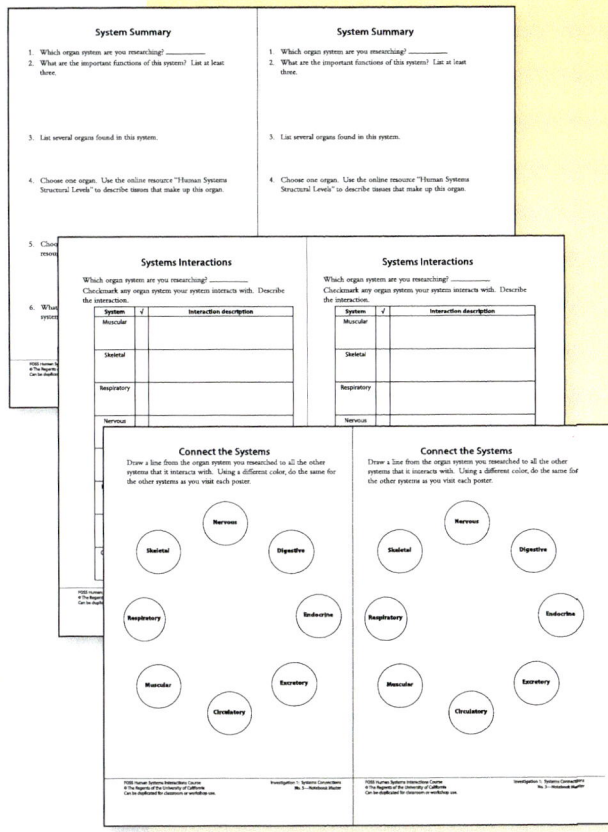

Nos. 1–3—Notebook Masters

INVESTIGATION 1 – Systems Connections

Day 1: Organ-system research

Day 2: System-interactions research using computers

Day 3: Disease research and preliminary diagnosis

Day 4: *Doctor Interview 2* (and final diagnosis)

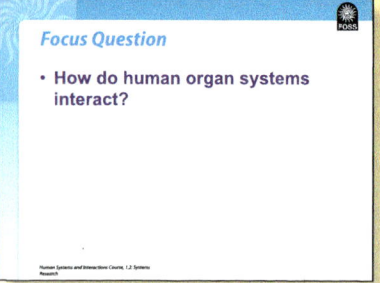

Teaching slides, 1.2

GETTING READY for
Part 2: Systems Research

Quick Start

Schedule	• Computers with Internet access, one for each pair of students 4 sessions including research and video
Preview	• Preview the FOSSweb Resources by Investigation for this part (such as printable masters, teaching slides, and multimedia) • Preview the reading: "Human Organ Systems," Step 6 • Preview the online activity: "Human Systems Structural Levels," Step 8 • Preview the video: *Doctor Interview 2*, Step 22
Print or Copy	**For each student** • Notebook sheets 1–3 **For the teacher** • *Performance Assessment Checklist*
Prepare Material	• Organize locations for systems posters **A** • Obtain mini-whiteboards and marking pens **B**
Plan for Assessment	• Review Step 19, "What to Look For" in the performance assessment **C**

▶ **NOTE**
Notebook sheets are available on FOSSweb in several formats. For each notebook sheet, you can select "to photocopy," which will be identical to the printed notebook masters in **Teacher Resources**, or "for display," which is rotated and zoomed for easier display. You can also type into these notebook sheets while projecting them.

84   Full Option Science System

Part 2: Systems Research

Preparation Details

A. Organize locations for systems posters

In Step 12, groups summarize how the organ system they are researching interacts with other systems. They write connections on self-stick notes and place them on posters of the other systems. Use painter's tape to attach the posters to classroom walls, spread around the room for easy access. In Step 14, students use colored pencils to connect systems to other systems on notebook sheet 3, You may want to put four pencils of the same color at each poster for each group to use.

> **NOTE**
> The class will need colored pencils— four each of eight different colors.

B. Obtain mini-whiteboards and marking pens

As an option in Step 2, you might want to provide mini-whiteboards, whiteboard marking pens, and erasers for groups to list human organ systems. If you cannot obtain mini-whiteboards, consider placing a piece of white card stock in a plastic sheet protector as an alternative.

C. Plan for performance assessment

Assessing the three dimensions envisioned in the NRC Framework and the NGSS performance expectations requires you to peek over students' shoulders while they are in the act of doing science or engineering. Observing the rich conversation among students and the actions they are taking to investigate phenomena or design solutions to problems can provide important information about student progress.

In Step 19, students engage in argument from evidence about the diagnosis of a patient's disease. This is an opportunity to observe students' use of science practices, content understanding, and crosscutting concepts. Preview What to Look For in Step 20 of Guiding the Investigation.

Photocopy the *Performance Assessment Checklist* for this investigation part. You will need one copy for each class. There are two forms of the checklist available, one for group assessment and one for individual student assessment. You can also download the checklist as an electronic spreadsheet found on FOSSweb in teacher resources. You will need to enter your students' names or the group names on the sheet before using it.

Carry the checklist with you as you observe and listen to students' discussions and observe their interactions. For more information about how to conduct performance assessments, see the Assessment chapter.

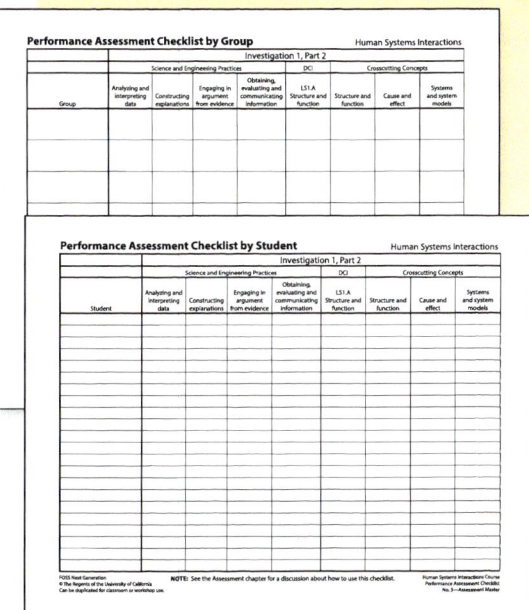

Performance Assessment Checklists

Human Systems Interactions Course—FOSS Next Generation

85

INVESTIGATION 1 – Systems Connections

FOCUS QUESTION

How do human organ systems interact?

GUIDING *the Investigation*
Part 2: *Systems Research*

SESSION 1

Students will . . .
- Review patient's symptoms (Step 1)
- Make a list of human organ systems (Step 2)
- Focus on one organ system to assist in diagnosis (Steps 3–5)
- Read "Human Organ Systems," focusing on one system (Steps 6, 7)
- Use online resources to gather information on organ systems (Steps 8, 9)

SESSION 2

Students will . . .
- Use notebook sheet 2, *Systems Interactions*, to organize data (Steps 10, 11)
- Share results of organ-system interactions using posters (Steps 12, 13)
- Summarize the analysis of system interactions and discuss example of diabetes (Steps 14, 15)

SESSION 3

Students will . . .
- Discuss how patient's symptoms affect organ systems (Step 16)
- Gather more information about patient's symptoms (Step 17)
- Research three suspected diseases using "Disease Information" (Step 18)
- Engage in argumentation from evidence concerning the diagnosis (Steps 19, 20)
- Review vocabulary and answer the focus question (Steps 21, 22)

SESSION 4

Students will . . .
- View video revealing the doctor's diagnosis of the patient (Step 23)
- Review notebook entries for the investigation and develop a list of key points in groups and as a class (Step 24)

Part 2: Systems Research

SESSION 1 45–50 minutes

1. **Review symptoms**

   Refer to the list of patient symptoms. Tell students that they will continue their investigation into those symptoms, but they have to gather more information about the human body to understand how the symptoms relate to the various organ systems. Explain,

   Another complication in this research is that some organ systems might start malfunctioning and affect other organ systems. We need to understand what the structural levels of each system are and then research how those systems interact to see if one system's breakdown could contribute to the breakdown of others.

2. **List human organ systems**

   Students may not remember the major organ systems that make up the human body. Ask them to recall the structural levels (or levels of complexity). Project the Organ System card and have students discuss with a partner what they recognize about the system illustrated. Ask,

   ➤ *What structures do you recognize in this illustration?* [Heart, veins, arteries.]

   ➤ *These structures are organs. What is their function?* [To work together to circulate blood throughout the body.]

   ➤ *These organs belong to one organ system. What is it called?* [**Circulatory system**.]

   Give groups a few minutes to work together to come up with any other systems that they can think of. They can list their responses on a mini-whiteboard. Cruise around the room to check on progress and ask the following questions if groups get stumped.

   ➤ *What organs does the body use in order to breathe? What is that system called?* [Lungs, nose, mouth; **respiratory system**.]

   ➤ *What happens to the food we eat? What system takes care of processing food?* [The food we eat gets digested; **digestive system**.]

   ➤ *How does our body gather information about the environment?* [Through the five senses (the **nervous system**).]

   ➤ *What changes dramatically when humans are adolescents?* [Hormones (the **endocrine system**).]

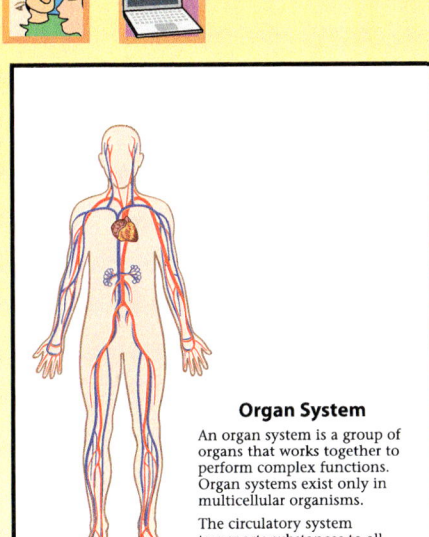

Human Systems Interactions Course—FOSS Next Generation

87

INVESTIGATION 1 – Systems Connections

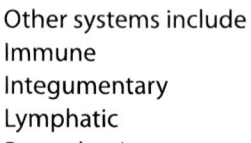

Organ Systems

Students will research these eight systems.
Circulatory
Digestive
Endocrine
Excretory
Muscular
Nervous
Respiratory
Skeletal

Other systems include
Immune
Integumentary
Lymphatic
Reproductive

Ask one group at a time to suggest an organ system. Write the names of the systems on chart paper or on the board. (If you want a permanent record, use chart paper.) Continue the list until the class exhausts their current knowledge, and then add to it if necessary.

3. Gather information to make a diagnosis
Ask,

➤ *We have a list of organ systems here. Which ones are you familiar with and which ones do you want to know more about? What questions do you have about these systems?*

Look for student questions about how the systems interact. Ask,

➤ *What information about the organ systems do we need in order to diagnose a patient's disease?*

Ask groups to turn in their notebooks to the list of patient symptoms presented in the *Doctor Interview 1* video. Have students review these symptoms in their groups as they consider what they need to know about the human organ systems.

4. Focus question: How do human organ systems interact?
Tell students that the first thing they need to do is to find out more about the organ systems themselves and then think about how the systems interact. They might use that information to diagnose the patient's illness. Introduce the focus question. Project or write it on the board. Ask students to turn in their notebooks to a clean page and write the focus question at the top.

➤ *How do human organ systems interact?*

Students should leave space in their notebooks so they can answer the focus question at the end of their investigation.

5. Introduce researching organ systems
Tell students that each group will research one organ system, explore how that system interacts with other systems, and then share their findings with the class. It is important that they do a thorough job and be prepared to communicate what they learn. Assign each group an organ system. Use a method that you feel comfortable with: allow groups to choose, or assign them randomly. Ideally, students should pick a system that interests them or with which they are not already familiar.

Distribute a copy of notebook sheet 1, *System Summary,* to each student. Tell students that while their group will work together, each student is responsible for recording the findings.

ELA CONNECTION

This suggested strategy addresses the Common Core State Standards for ELA for literacy.

WHST 8: Gather relevant information from multiple print and digital sources.

88   Full Option Science System

Part 2: Systems Research

READING *in Science Resources*

6. Read "Human Organ Systems"

Distribute a copy of *FOSS Science Resources: Human Systems Interactions* to each student. If you haven't already done so, introduce *Science Resources*. Give students a few minutes to look at and discuss the cover of the book, and to examine and discuss the table of contents. Point out that the book has readings in the front and data in the back. Students should also locate the glossary and the index.

Have students turn to the article "Human Organ Systems" and give them a moment to discuss the title, photographs, and text features on the first pages.

7. Use a reading comprehension strategy

Read the introduction on page 3 aloud or call on a volunteer. Explain that they will be doing a jigsaw of the reading. Each group will become an expert on their assigned organ system and will use the reading and the online activity to gather relevant information about how the system interacts with other systems. Each group will then share the systems interaction with the rest of the class.

Suggest that students follow this procedure working in their groups:

a. Preview the text by reading the titles and captions and looking over the words in bold. Then, discuss the images with a partner. When looking at the photographs ask, "What do we notice? What is happening?" When looking at the diagrams ask, "What information does this diagram present? What do we already know about this? What do the different colors mean? What do the lines and arrows mean? (The diagram of the digestive system on page 13 of the *FOSS Science Resources* book is a good example if you want to model this strategy with the whole class or small group.)

b. Read the article independently. Place self-stick notes on places in the text you think are important and interesting, and where you have questions or find the text confusing.

c. Discuss the text with your group. Take turns sharing the new information you learned and review any questions, new words or confusing parts you encountered.

d. Review the *System Summary* sheet and assign sections to each member of the group to address. Reread the text or explore the online resource to find and record the answers. (Students will need to go online to FOSSweb in Step 8 to answer questions 4 and 5.)

e. Share out responses and discuss any questions or ideas.

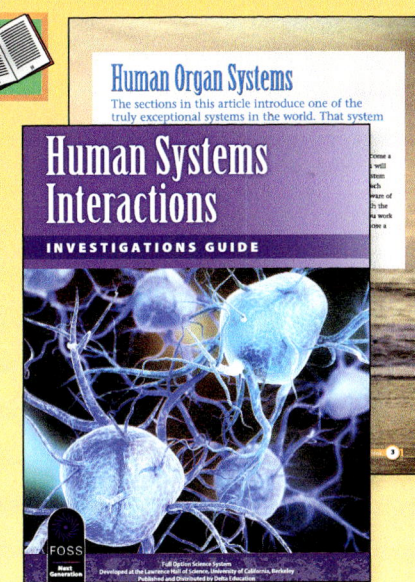

FOSS Science Resources

ELA CONNECTION

These suggested strategies address the Common Core State Standards for ELA for literacy.

RST 4: Determine the meaning of symbols, key terms, and other domain-specific words and phrases.

RST 5: Analyze the structure an author uses to organize a text.

RST 7: Integrate quantitative or technical information expressed in words with visual representation.

RST 10: Read and comprehend science/technical texts independently and proficiently.

SL1: Engage in collaborative discussions.

SCIENCE AND ENGINEERING PRACTICES

Obtaining, evaluating, and communicating information

INVESTIGATION 1 – Systems Connections

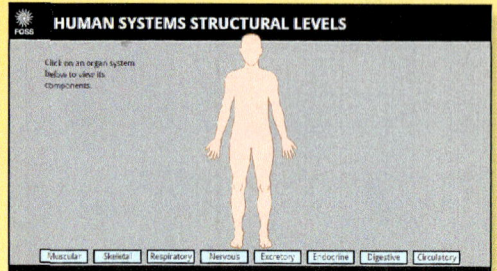

8. Introduce online activity

Distribute a laptop or tablet to each group and show them how to access the online activity.

Project the online "Human Body Structural Levels" activity. Demonstrate how to click a system image to access more information about that system. Clicking a red highlighted section will lead to the next structural level. Students will find information to answer items 4 and 5 on notebook sheet 1.

9. Start research

Give groups the rest of the session to work on the notebook sheet for their system. As students are working, cruise the groups and check that they are recording accurate information.

SESSION 2   *45–50 minutes*

SCIENCE AND ENGINEERING PRACTICES

Constructing explanations

Obtaining, evaluating, and communicating information

CROSSCUTTING CONCEPTS

Systems and system models

Structure and function

10. Research systems interactions

Before describing the next part of the research, determine where the groups are. When the first groups appear to be finishing notebook sheet 1, introduce notebook sheet 2, *Systems Interactions*. Tell students that when they complete the first part of the research, they can look at how their organ system interacts with other systems in the body.

Project a copy of notebook sheet 2, *Systems Interactions,* and tell students to fill in the information using the descriptive data from the *FOSS Science Resources* book.

If necessary, demonstrate how students will fill in the table. The "description" column should be a list of the points of interaction with your organ system. Fill in the name of an organ system and write a sample description for the "muscular" row. You can use the first part of the answer sheet for this sample.

Systems Interactions

Which organ system are you researching? __skeletal__

Checkmark any organ system your system interacts with. Describe the interaction.

System	√	Interaction description
Muscular	√	Bones are attachment points for muscles to help body move.
Skeletal		

90   Full Option Science System

Part 2: Systems Research

11. Record interactions

Give students time to research interactions. Remind them to use their *FOSS Science Resources* as a reference. Again, cruise the groups to see that students are recording systems interactions on notebook sheet 2.

12. Post organ-systems interactions

When most groups are done, use the example in the sidebar to illustrate how students will record their information on self-stick notes. Write the example on the board as you explain.

When you complete your interactions research, ask a Getter to obtain one self-stick note for each system you checked off on notebook sheet 2.

*Write your organ system and one other organ system it interacts with, and then write how the systems interact. For instance, my group researched the **skeletal system**. So we write "Skeletal System" on top of the self-stick note. The skeletal system interacts with the **muscular system**, so we write a slash followed by "Muscular System." We write how the skeletal system interacts with the muscular system. Then we put that self-stick note on the Muscular System poster (not on the Skeletal System poster). Prepare and post one self-stick note for each organ system your organ system interacts with.*

> Skeletal System/
> Muscular System
>
> Skeletal muscles are attached to bones.

Show students where the self-stick notes are and where each poster is located, and let them get back to work.

Again, cruise the groups to see that students are correctly transferring interactions to self-stick notes and posting them on the appropriate posters. For instance, the skeletal system group should have a second sticky note indicating a relationship with the circulatory system directly and digestive, respiratory, and excretory systems indirectly because bone is living tissue.

13. Share results

When each group has posted its organ system's interactions on the posters around the room, give the groups a few minutes to move to and look at the poster that represents their *own* system. They can add to notebook sheet 2 or another page in their notebook, using a different-colored pencil to indicate that the new information comes from a different source.

TEACHING NOTE

If it is more convenient, ask groups to retrieve their own system poster and take it back to their table. When groups "do the rounds" in the next step, they can rotate around the tables instead of the walls.

Call students back to their seats and pose these questions for them to discuss with their table groups.

➤ What did you find that was surprising to you?

➤ What do you observe about how the organ system you researched interacts with the other systems in the body?

Call on Reporters to share their group's observations.

Human Systems Interactions Course—FOSS Next Generation

INVESTIGATION 1 – Systems Connections

14. Compile whole-class system interactions

Tell students that it is time to take a look at all the posters. Distribute a copy of notebook sheet 3, *Connect the Systems*, to each student. Project a copy of the notebook sheet and demonstrate how to fill it in. Students should use a different color of pencil for each system. (Place four colored pencils of the same color by each poster for students to use.) They will start with the system that they researched and draw a line between that system and the others that it interacts with, as indicated by the self-stick notes on the poster. They can use rulers if they like.

When you call time, the groups will rotate to the next system poster and draw different-colored lines between that system and the ones it interacts with. Soon, students may just start drawing lines, but encourage them to look for the evidence that other groups wrote on the self-stick notes. This step should move fairly quickly.

After groups have visited each poster, call them back to their seats and discuss their observations. Ask,

➤ *What do you observe about how the organ systems in the body interact with the other systems in the body?* [Most of the organ systems interact with all of the others.]

➤ *What might happen if one system is affected by a disease or condition and doesn't function?* [Other systems could be affected in one way or another.]

15. Discuss diabetes example

Give students the example of diabetes. Project page 97 of the *FOSS Science Resources* book, "Diabetes Affects Human Organ Systems."

Some students may have personal or familial experience with diabetes. If there is time, ask students to share their experience. Listen as the discussion unfolds. Then say,

*A major symptom of **diabetes** is the buildup of sugar in the blood. The pancreas is an organ that produces the **hormone** insulin, which normally regulates the level of sugar.*

➤ *Which group researched the system that produces hormones?* [Endocrine group.]

➤ *Tell us what you found out about insulin.* [Insulin controls the cell's ability to absorb sugars from the bloodstream.]

In diabetes, either your insulin-producing cells no longer work, or the target cells in your body no longer recognize insulin. Either way, your sugar-regulating system is out of whack. The kidneys, which work to filter excess sugar out of the blood, can't keep up, and the extra sugar is eliminated in the urine.

TEACHING NOTE

You can have groups take the self-stick notes from their own poster and divide them among the members of the group to keep in their notebooks.

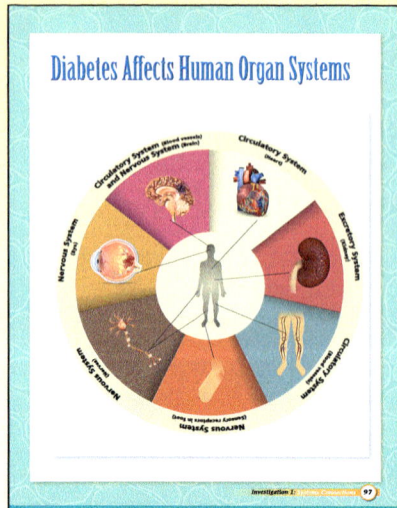

FOSS Science Resources

92   Full Option Science System

Part 2: Systems Research

➤ Which system includes the kidneys? [**Excretory**.]

With the increased urination, you feel thirsty, drink more, and consequently continue to urinate. A high level of sugar pulls fluids from your tissues, including the lenses of your eyes, leading to blurred vision. Excess sugar also can lead to nerve damage, either decreasing sensation or causing pain in the hands and feet.

➤ Which system includes vision and the senses? [Nervous.]

Diabetes is an example of a disease that starts with decline of one organ system which leads to decline in other organ systems.

CROSSCUTTING CONCEPTS

Cause and effect
Systems and system models
Structure and function

SESSION 3 45–50 minutes

16. Discuss symptoms
Refer to the patient's symptoms. Tell students that now they have enough background about the body to look more closely at the diseases that might be affecting the patient. Give groups a few minutes to consider the following question.

➤ *Think about the symptoms of the patient we are trying to diagnose. Do they affect any of the organ systems you have researched?*

Write down student ideas next to each symptom.
- Fever [Don't know, maybe all.]
- Fatigue (tiredness) [Don't know, maybe all.]
- Muscle aches (especially back) [Muscular.]
- Joint pain [Skeletal.]
- Severe headaches [Nervous.]

TEACHING NOTE

Fever is actually a function of the immune system as it fights infection. Fever is also controlled by the hypothalamus in the brain, which acts as the body's thermostat.

17. Disclose more symptoms
Tell students that the patient shared three more symptoms:
- chest pain
- shortness of breath
- weird feeling in the heart

Suggest that students add the symptoms to the list in their notebooks that they started in Part 1 after viewing the video. They should think about which other systems may be affected.

Human Systems Interactions Course—FOSS Next Generation

INVESTIGATION 1 – Systems Connections

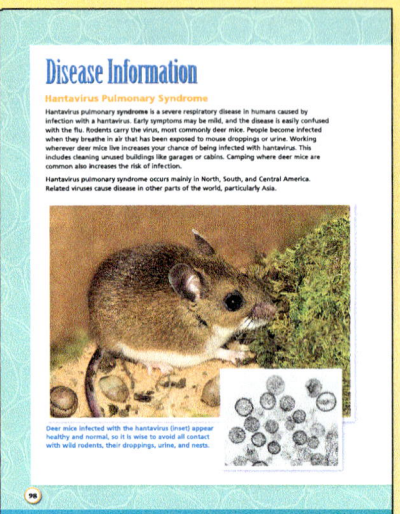

FOSS Science Resources

E L N O T E
Suggest students use a graphic organizer to list the symptoms for each disease.

E L N O T E
For students who need support, provide sentence frames and prompts such as
"I agree/disagree with _____ because _____."
"What is the evidence that _____?"
"Can you say more about _____?"
"Have you considered _____?"
"I think _____ supports the claim that _____."

READING in Science Resources

18. Research the suspected diseases

Have students turn to "Disease Information" in *FOSS Science Resources*. Suggest to students that it may be time to look a little more closely at the three diseases the doctor suspects. Tell them to think about what more they need to know and how they should organize the information in their notebook. Make sure they include their findings and questions.

➤ *What information do you still need to help find out what disease is affecting the patient, so the doctor can begin treatment? What questions do you suggest the doctor ask the patient?*

Allow enough time for groups to read "Disease Information," record their findings and questions, and discuss the information as a group. Below is an example of a graphic organizer the students might use in their notebooks.

Patient's Symptoms	Hantavirus Pulmonary Syndrome	Lupus	Lyme Disease

19. Engage in argumentation

Tell students that they will work together with the rest of the class to make a recommendation to the doctor about what they think is causing the patient's symptoms.

➤ *What disease do you think is causing the patient's symptoms?*

Give students a few moments in their groups to prepare their response and list any questions they want to ask the patient. Prepare to record each group's information and questions and share them via projection or on chart paper.

Part 2: Systems Research

If this is the first time students have presented arguments (claims based on evidence) you might want to spend a few minutes discussing expectations and how you want students to interact. Note that groups may disagree on the disease affecting the patient. If necessary, review norms for collaborative discussions as described on page 9 in the Science-Centered Language Development Chapter in *Teacher Resources*. Tell students to use evidence from the readings and reasoning to support their claims and to counter the claims they disagree with.

NOTE: It is important that you do not tell students the final diagnosis on this day or even the next. The next day, students will watch the second part of the video (see Step 22). There is plenty of evidence in this video for students to identify the disease affecting the patient as hantavirus pulmonary syndrome. Students should come to the conclusion on their own.

20. **Assess progress: performance assessment**

    As students make their arguments, note their progress on the *Performance Assessment Checklist*.

    What to Look For

    - *Students effectively communicate a claim based on evidence. (Analyzing and interpreting data; constructing explanations.)*
    - *Students engage in argumentation related to the symptoms and information they have gathered regarding the patient. (Engaging in argument from evidence; LS1.A: Structure and function.)*
    - *Students effectively argue for the disease they think is affecting the patient's organ systems; arguments are backed by supportive evidence including symptoms and research. (Engaging in argument from evidence; structure and function; cause and effect; systems and system models.)*
    - *Students communicate and listen to each other's arguments and respond with their own evidence. (Obtaining, evaluating, and communicating information.)*

    As you monitor the students, note any alternate conceptions they might have and determine the feedback or next-step strategy you plan to use moving forward.

ELA CONNECTION

These suggested strategies address the Common Core State Standards for ELA for literacy.

RST 1: Cite evidence to support analysis of science texts.

SL1: Engage in collaborative discussions.

SL4: Present claims and findings.

SCIENCE AND ENGINEERING PRACTICES

Analyzing and interpreting data

Constructing explanations

Engaging in argument from evidence

Obtaining, evaluating, and communicating information

DISCIPLINARY CORE IDEAS

LS1.A: Structure and function

CROSSCUTTING CONCEPTS

Cause and effect

Systems and system models

Human Systems Interactions Course—FOSS Next Generation

INVESTIGATION 1 – Systems Connections

circulatory system
diabetes
digestive system
endocrine system
excretory system
hormone
muscular system
nervous system
respiratory system
skeletal system

21. Review vocabulary

Take a few moments to review the vocabulary developed in this investigation. Ask students to highlight vocabulary words and definitions in their notebooks and to update their vocabulary indexes.

22. Answer the focus question

Ask students to return to the focus question in their notebooks.

➤ *How do human organ systems interact?*

Ask students to write their best response, based on the work they have done. They should include how organ systems interact, citing examples; how diseases may affect different systems; and the effect a diseased organ system might have on other systems. Give students a few minutes to share their responses with a partner and revise if needed.

Tell students you will show the final doctor interview the next session after all classes have shared their diagnoses.

SESSION 4 *45–50 minutes*

23. View video

When all classes have made their tentative diagnoses, watch *Doctor Interview 2*. Tell students to look and listen for results of tests and other information the doctor found out about the patient. When the video is done, give students a few more minutes with the "Disease Information" data to either change or add to their first diagnosis.

Open the discussion to the entire class. Again, look for students basing their claims on evidence. Encourage students to listen carefully to each other's arguments and to evaluate the soundness of the reasoning and relevance and sufficiency of the evidence.

ELA CONNECTION

This suggested strategy addresses the Common Core State Standards for ELA for literacy.

SL 3: Delineate and evaluate a speaker's argument.

Part 2: Systems Research

WRAP-UP

24. Review notebook entries

Ask students to go through their notebook entries and select one big idea that summarizes an important finding from this investigation. They should record this point in their notebooks.

Have students share their big ideas with their group, selecting one big idea within the group to share with the class. Create a class chart of big ideas by recording each group's idea on a piece of chart paper, whiteboard, or document projected from the computer. You might need to help groups rephrase their big ideas for clarity.

Students should record the big ideas in their notebooks and reference the page numbers in their notebooks where additional information supports each one. Here are the big ideas that should come forward from Investigation 1.

- Humans are multicellular organisms composed of organ systems which are made of organs, which are made of tissues, which are made of cells.
- Cells are composed of cell structures (organelles), which are made of molecules, which are made of atoms.
- The human body is a system of interacting subsystems (circulatory, digestive, endocrine, excretory, muscular, nervous, respiratory, and skeletal).
- When one organ system is compromised, it affects other organ systems. Diagnosing disease requires understanding how systems interact.

DISCIPLINARY CORE IDEAS

LS1.A: Structure and function

INVESTIGATION 1 – Systems Connections

EXTENDING *the Investigation*

- **Research diseases**

    Have students select a medical disease or condition that they are interested in and find out more about it. A good trusted source of reliable and appropriate online information is available from the US National Library of Medicine and the National Institutes of Health—MedlinePlus.

    This government site includes a medical dictionary and encyclopedia, extensive information on over 975 diseases and conditions, health information in Spanish, and is appropriate for students and families. There is no advertising on this site, nor does MedlinePlus endorse any company or product.

- **Talk to a doctor**

    Have students talk to their primary-care physicians about any patient cases that may have stumped them. How did they reach a diagnosis?

- **Research other human organ systems**

    Have students research other systems such as immune, integumentary, lymphatic, and/or reproductive.

- **Research standard first aid**

    First aid is assistance provided to a person that has an injury or sudden illness in order to save a life or stabilize the condition until medical help arrives. First aid can also be treating a minor condition due to an injury. Find out about some standard first aid treatments for problems or take a first aid course.

INVESTIGATION 2 – Supporting Cells

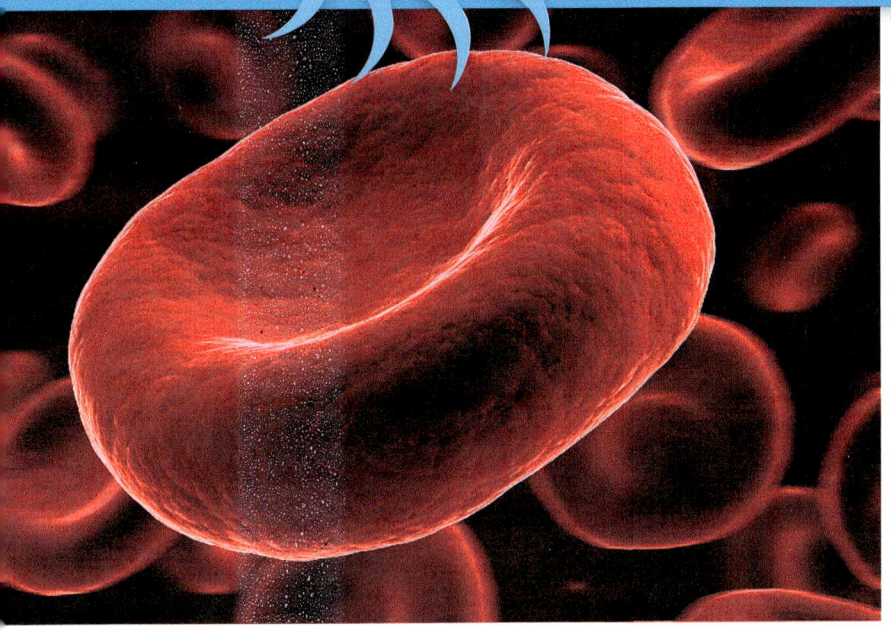

Part 1	
Food and Oxygen	110
Part 2	
Aerobic Cellular Respiration	123

PURPOSE

In *Supporting Cells,* students fatigue their muscles and learn how their cells obtain the food and oxygen they need via the digestive, respiratory, and circulatory systems. They model how aerobic cellular respiration works in cells. They find out how cells eliminate the wastes produced during aerobic cellular respiration via the circulatory, respiratory, and excretory systems.

Content

- The human body is a system of interacting subsystems.
- The respiratory system supplies oxygen and the digestive system supplies energy (food) to the cells in the body.
- The circulatory system transports food and oxygen to the cells in the body and carries waste products to the excretory/respiratory systems for disposal.
- Aerobic cellular respiration is the process by which energy stored in food molecules is converted into energy for cells.

Practices

- Develop models to describe how food molecules are rearranged by chemical reactions forming new molecules to provide usable energy for cells.
- Construct explanations about organ system interactions at different scales.

Science and Engineering Practices
- Developing and using models
- Constructing explanations
- Obtaining, evaluating, and communicating information

Disciplinary Core Ideas
PS3: How is energy transferred and conserved?
PS3.D: Energy in chemical processes and everyday life
LS1: How do organisms live, grow, respond to their environment, and reproduce?
LS1.A: Structure and function
LS1.C: Organization for matter and energy flow in organisms

Crosscutting Concepts
- Scale, proportion, and quantity
- Systems and system models
- Energy and matter

Full Option Science System

INVESTIGATION 2 – *Supporting Cells*

	Investigation Summary	Time	Focus Questions and Practices
PART 1	**Food and Oxygen** Students participate in an exercise activity to think about how the cells in the human body get oxygen and energy (food). They watch video clips and manipulate an online activity to add detail to their ideas. They construct a model to illustrate the pathways that oxygen and energy (food) take from the external environment to a muscle cell in the leg.	**Active Inv.** 3 Sessions	**How do cells in the human body get the resources they need?** **Practices** Developing and using models Constructing explanations Obtaining, evaluating, and communicating information
PART 2	**Aerobic Cellular Respiration** Students model the substances and steps in aerobic cellular respiration. They summarize the entire process, demonstrating how substances get *to* the cells, what happens *at* the cells, and how substances depart *from* the cells to be removed from the body.	**Active Inv.** 2 Sessions **Assessment** 2 Sessions	**How does the energy in food become energy that cells can use?** **Practices** Developing and using models Constructing explanations Obtaining, evaluating, and communicating information

* A class session is 45–50 minutes.

At a Glance

Content Related to DCIs	Literacy/Technology	Assessment
• The human body is a system of interacting subsystems. • The respiratory system supplies oxygen and the digestive system supplies energy (food) to the cells in the body. • The circulatory system carries food and oxygen to the cells in the body and carries waste products to the excretory/respiratory systems for disposal.	**Science Notebook Entry** *Organ-Systems Video Questions* **Online Activity** *"Human Cardiovascular System"* **Videos** *Digestive and Excretory Systems* *Circulatory and Respiratory Systems*	**Embedded Assessment** Response sheet
• The human body is a system of interacting subsystems. • The respiratory system supplies oxygen and the digestive system supplies energy (food) to all the cells in the body. • The circulatory system carries food and oxygen to all the cells in the body and carries waste products to the excretory/respiratory systems for disposal. • Aerobic cellular respiration is the process by which energy stored in food molecules is converted into usable energy for cell.	**Science Notebook Entry** *To and From the Cells* **Science Resources Book** *"Aerobic Cellular Respiration"* **Video** *Digestive and Excretory Systems*	**Embedded Assessment** Performance assessment **Benchmark Assessment** *Investigations 1–2 I-Check* **NGSS Performance Expectations addressed in this investigation** MS-LS1-3 MS-LS1-7

Human Systems Interactions Course—FOSS Next Generation   **101**

INVESTIGATION 2 – Supporting Cells

SCIENTIFIC *and Historical Background*

How Do Cells in the Human Body Get the Resources They Need?

Run up a steep hill. Swim a lap as fast as you can. Sprint on a bike. Why do you get tired? Your body needs energy to exercise, and you get tired because you are using up your body's available energy supplies. How

© Maridav/Shutterstock

can you get more energy? Most athletes will tell you that you need to keep eating to replenish your energy supplies. You cannot regain expended energy until you eat food.

Energy. Energy comes from food. Food molecules such as carbohydrates, fats, and proteins have potential energy in their bonds. After those bonds break and rearrange, they release energy. The amount of energy released depends on the type of molecule, but it is always measured in units of **calories**. A calorie is the amount of energy needed to raise the temperature of 1 gram (g) of water 1 degree Celsius (°C).

Your body needs water, but it does not provide energy as it produces no calories. What about sleep? Sleep helps your body function better, but again, it produces no calories and therefore no energy. And caffeine? Caffeine is a stimulant. It increases neural activity in your brain, which stimulates the pituitary gland to release hormones that signal the adrenal gland to produce adrenaline. Adrenaline travels throughout the body with numerous effects. However, the body cannot access calories from caffeine, hence no usable energy. The only energy the human cell can access is energy that comes from food molecules.

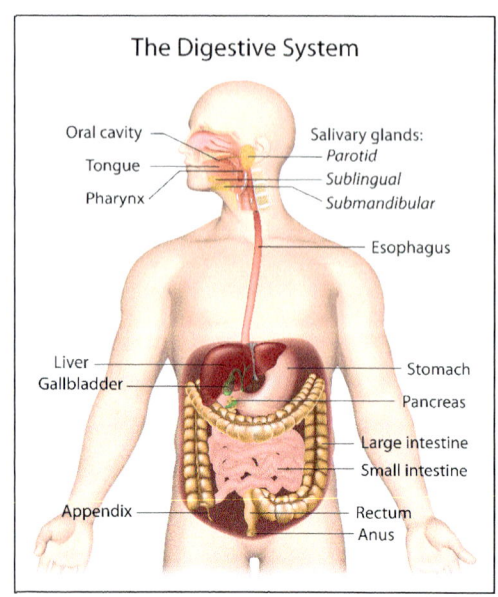
© alila/Shutterstock

A cell cannot eat a strawberry, but a strawberry contains energy stored in sugar molecules. The body has an elaborate system for breaking the strawberry down in order to make sugar molecules available to cells. That system is the digestive system. The digestive system chemically and mechanically breaks food into nutrient molecules. The nutrient molecules are absorbed through the villi in the wall of the small intestine into **capillaries**. From there, the bloodstream carries the nutrients to every cell in the body.

Thus the digestive system uses the circulatory system to deliver the goods, so to speak.

102   Full Option Science System

Scientific and Historical Background

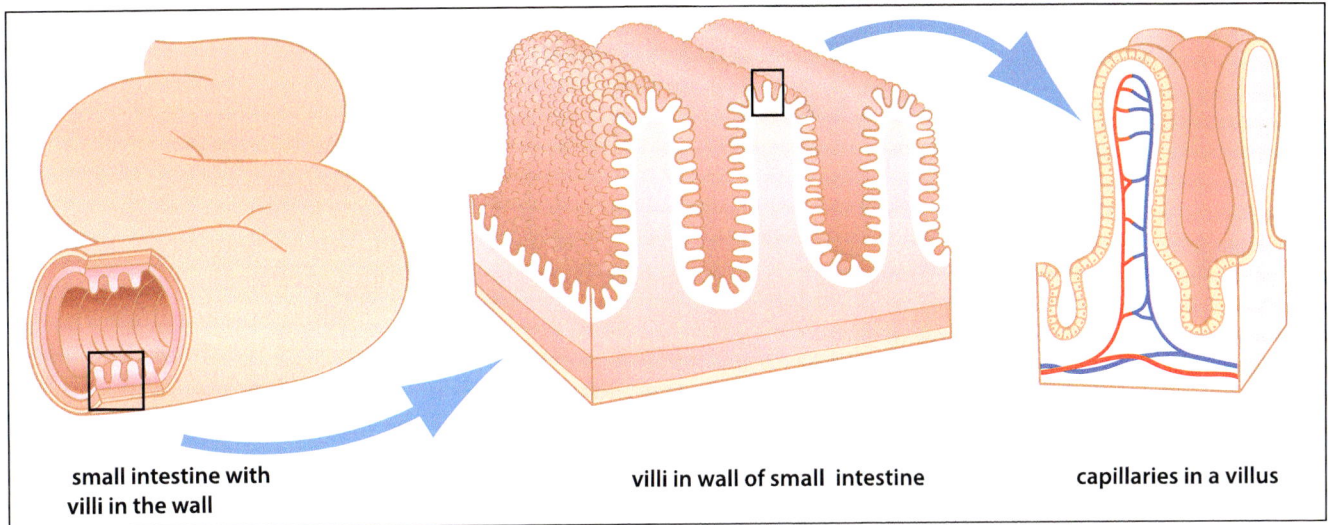

small intestine with villi in the wall

villi in wall of small intestine

capillaries in a villus

Oxygen. Our breathing rate is affected by heavy exercise. We breathe faster and deeper to bring as much air as possible into our body. The component of air that is most important for our cells is oxygen.

The respiratory system brings oxygen in from the environment. As the lungs inflate, oxygen (and other gases in air) floods through increasingly narrower tubes to the remotest recesses, ending in tiny air sacs called **alveoli** (singular, **alveolus**). In the alveoli, oxygen diffuses into capillaries, and from there, the bloodstream carries the oxygen molecules to cells in the body. Sound familiar? The respiratory system uses the circulatory system to deliver the oxygen to the cells.

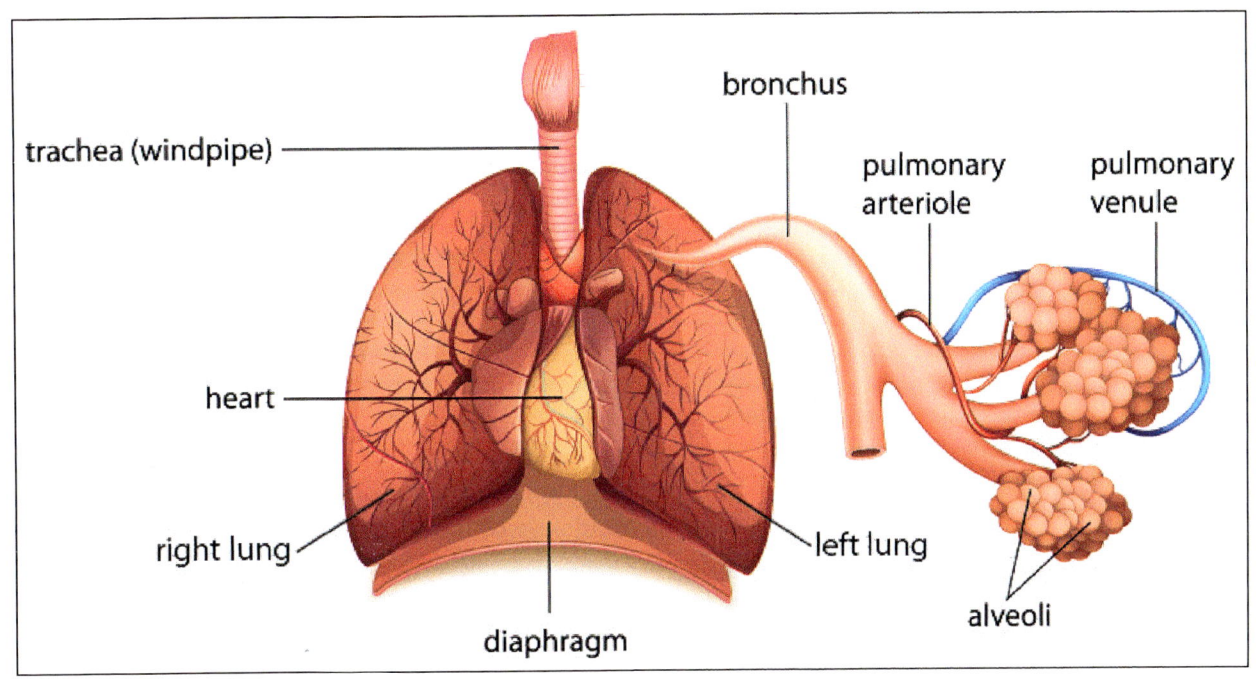

© BlueRingMedia/Shutterstock

Human Systems Interactions Course—FOSS Next Generation   **103**

INVESTIGATION 2 – Supporting Cells

How Does the Energy in Food Become Energy That Cells Can Use?

Nutrient molecules and oxygen are transported by the bloodstream to a cell. The molecules diffuse across the capillary walls and across the cell membrane into the interior of the cell. Diffusion is essentially a simple movement of molecules from a region of high concentration to a region of lower concentration. Large molecules can't simply pass through the cell membrane. Transport proteins embedded in the membrane facilitate the movement of larger nutrient molecules, and sometimes energy is needed for an extra boost to move nutrients into the cell. (This process is called active transport.) Once the food molecules are in the cell, the molecules combine with oxygen to release the energy stored in the nutrient molecule's chemical bonds. Most cells use simple sugars such as **glucose** as their primary energy source, so we will focus on that particular molecule.

If a cell actually burned the glucose molecules, the energy would be released as heat and light, a kind of energy that is not appropriate for the cell's use. Instead, the energy stored in the bonds of the glucose molecules must be transformed into a form of energy that the cell can use. This happens over a series of complicated chemical reactions. In summary, oxygen reacts with glucose to form carbon dioxide and water, transferring energy to an energy-rich molecule called adenosine triphosphate (ATP). ATP is then used to fuel metabolism and support growth and repair in cells.

$$C_6H_{12}O_6 + 6O_2 \longrightarrow 6CO_2 + 6H_2O + \text{ATP}$$
glucose   oxygen   carbon dioxide   water   energy

This process is called **aerobic cellular respiration**. Aerobic processes require oxygen. Cellular respiration refers to how cells get usable energy from food. (We normally equate respiration with breathing, but *cellular* respiration refers specifically to the energy-releasing pathways within the cell.)

Aerobic cellular respiration occurs in three stages—glycolysis, the Krebs Cycle, and the electron transport chain. The first stage of cellular respiration, glycolysis, occurs in the cytoplasm. Here, one 6-carbon molecule of glucose is converted into two 3-carbon molecules of pyruvic acid. Very little energy is released here; most of the energy stored in the bonds of the glucose molecule are still locked in the bonds of the pyruvic acid molecules. The pyruvic acid moves into the organelle called the mitochondrion, where the rest of the energy "extraction" transpires.

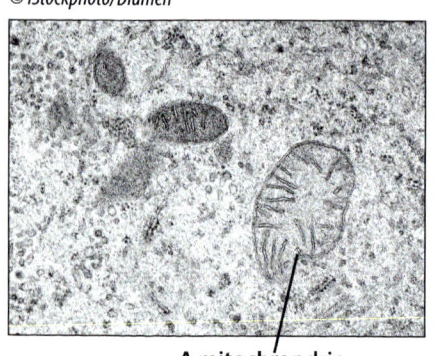

A mitochrondrion

104   Full Option Science System

Scientific and Historical Background

Mitochondria are very interesting on many levels. Mitochondria were once independent single-celled bacteria that, at some point in history, were engulfed by larger eukaryotic cells, resulting in symbiosis between the two. Over time, most of the mitochondrion's genome became incorporated into the genome of its host cells, and the mitochondrion's independence was lost. This happened early in the history of eukaryotic cells, and so all eukaryotic cells have mitochondria. Our cells could not generate the energy needed for life processes without mitochondria. Mitochondria are fascinating organelles, but most important for human life is their role in cellular respiration.

Once inside a mitochondrion, the pyruvic acid enters a complex cycle known as the Krebs Cycle, or the citric acid cycle. Here, the process becomes aerobic, using oxygen and producing carbon dioxide. A little more energy is released.

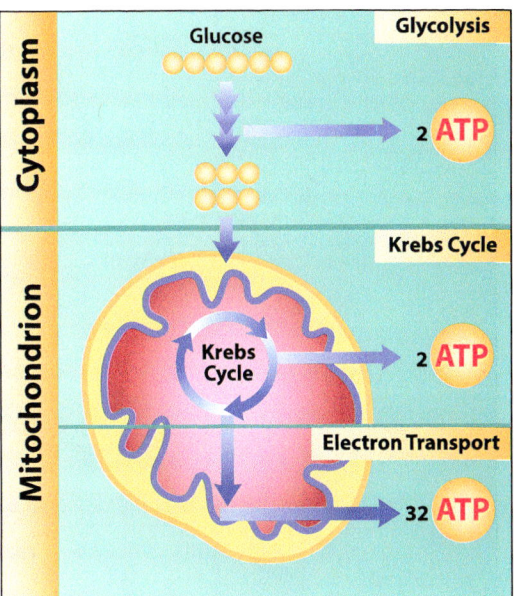

Most of the energy comes from the final stage of cellular respiration: the electron-transport chain. Oxygen is vital in this series of reactions. It is the final acceptor of electrons that are produced in glycolysis and the Krebs Cycle and passed down a chain of electron carriers in the inner membrane of a mitochondrion. Water is produced along the way, releasing a lot of energy. What's the important final product? A multitude of energy-rich molecules in the form of ATP that power the activities of the cell.

Waste. The other molecules, carbon dioxide and water, are considered waste. Oxygen and glucose were transported *to* the cells; *in* the cells, aerobic cellular respiration transferred the energy in the glucose bonds into energy-rich ATP molecules; and now, carbon dioxide and water must be transported *from* the cells and out of the body.

The waste molecules diffuse out of the cell into a neighboring capillary. The bloodstream carries them away. The carbon dioxide travels back to the lungs, where it moves into the alveoli and is expired. The water travels to the kidneys and is excreted.

Beyond glucose. Of course, glucose is not the only source of energy for our bodies. All complex carbohydrates we eat are broken down into sugars such as glucose, which start aerobic cellular respiration in the first stage. However, fats and proteins can be broken down into other molecules that enter cellular respiration at the Krebs Cycle. Fat is actually a more efficient fuel for the human body than glucose, and as we all know, the body stores extra energy as fat.

Human Systems Interactions Course—FOSS Next Generation

INVESTIGATION 2 – Supporting Cells

Organ-systems interactions. Aerobic cellular respiration is critical for generating energy for the body. It cannot happen without the necessary reactants. The organ systems must work together to deliver those molecules to the cells and remove the waste products of aerobic cellular respiration. The digestive system processes the food we eat and breaks it down into molecules such as glucose. The respiratory system brings in oxygen. The circulatory system carries both glucose and oxygen to the cells. And finally, the circulatory system carries the waste products of cellular respiration to the excretory and respiratory systems. If one part of the interaction falters, the cells will not produce the usable energy needed to continue life.

Teaching and Learning about Supporting Cells

TEACHING AND LEARNING *about Supporting Cells*

Developing Disciplinary Core Ideas (DCI)

"Ugh! I'm so hungry. I need to eat!" We hear this so often from our middle school students. And it's most likely true. Young people at this time in their lives are growing like weeds. And they need energy and building materials to grow. But all too often we see middle school students making nutritional choices that may be high in energy content, such as fats and sugar, but that may be terribly deficient in other essential nutrients for their bodies.

What students eat and drink affects how well their bodies function now and into the future. When students diet, they usually do not realize that depriving their bodies of food can lead to negative physical consequences (as well as associated psychological consequences). Nutritional deficiencies, growth deceleration, bone weakness, and in girls, menstrual irregularity are short-term as well as long-term risks. As students study the marvelous complexity of their bodies, they can learn to respect how finely tuned they are and how important it is to eat a healthy diet and to take care with their nutritional intake.

Students often confuse the intake of air through the mouth to the lungs (breathing of the respiratory system) with the chemical processes that produce usable energy in our cells (aerobic cellular respiration). Students need to understand that these processes are related but involve descriptions at different scales.

The experiences students have in this investigation contribute to the disciplinary core ideas **PS3.D: Energy in chemical processes and everyday life**; **LS1.A: Structure and function**; and **LS1.C: Organization for matter and energy flow in organisms**.

Engaging in Science and Engineering Practices (SEP)

In this investigation, students engage in these practices.

- **Developing and using models** about the interactions of organ systems at the organ level and cellular level to look at inputs and outputs.
- **Constructing explanations** using scientific ideas to describe how cells in the human body get the resources they need to survive.
- **Obtaining, evaluating, and communicating information** from books and media to construct explanations about the process of aerobic cellular respiration and communicate understandings.

NGSS Foundation Box for DCI

PS3.D: Energy in chemical processes and everyday life
- Cellular respiration in plants and animals involves chemical reactions with oxygen that release stored energy. In these processes, complex molecules containing carbon react with oxygen to produce carbon dioxide and other materials.

LS1.A: Structure and function
- In multicellular organisms, the body is a system of multiple interacting subsystems. These subsystems are groups of cells that work together to form tissues and organs that are specialized for particular body functions.

LS1.C: Organization for matter and energy flow in organisms
- Within individual organisms, food moves through a series of chemical reactions in which it is broken down and rearranged to form new molecules, to support growth, or to release energy.

NGSS Foundation Box for SEP

- **Develop and/or use a model to generate data** to test ideas about phenomena in natural or designed systems, including those representing inputs and outputs, and those at unobservable scales.
- **Apply scientific ideas, principles, and/or evidence** to construct, revise, and/or use an explanation for real-world phenomena, examples, or events.
- **Critically read scientific texts** adapted for classroom use to determine the central ideas and/or obtain scientific and/or technical information to describe patterns in and/or evidence about the natural and designed world(s).
- **Communicate scientific and/or technical information** in writing and/or through oral presentations.

Human Systems Interactions Course—FOSS Next Generation 107

INVESTIGATION 2 – *Supporting Cells*

NGSS Foundation Box for CC

- **Scale, proportion, and quantity:** The observed function of natural and designed systems may change with scale. Phenomena that can be observed at one scale may not be observable at another scale.
- **Systems and system models:** Systems may interact with other systems; they may have subsystems and be a part of larger complex systems.
- **Energy and matter:** Within a natural (or designed system), the transfer of energy drives the motion and/or cycling of matter. The transfer of energy can be tracked as energy flows through a designed or natural system.

Exposing Crosscutting Concepts (CC)

In this investigation, the focus is on these crosscutting concepts.

- **Scale, proportion, and quantity.** The function of the organ system can be described at the organ level and at the cell level.
- **Systems and system models.** Organ systems interact to perform function for the survival of the organism.
- **Energy and matter.** The transfer of energy can be traced from food produced by plants through the organ systems of the human body to provide glucose to the cell where the cells produce usable energy for cell processes.

Connections to the Nature of Science

- **Scientific knowledge assumes an order and consistency in natural systems.** Science assumes that objects and events in natural systems occur in consistent patterns that are understandable through measurement and observation. Science carefully considers and evaluates anomalies in data and evidence.
- **Science is a way of knowing.** Science is both a body of knowledge and processes and practices used to add to that body of knowledge. Scientific knowledge is cumulative and many people from many generations and nations have contributed to scientific knowledge. Science is a way of knowing used by many people, not just scientists.

Vocabulary
Aerobic cellular respiration
Alveolus
Calorie
Capillary
Glucose

Teaching and Learning about Supporting Cells

Conceptual Flow

The conceptual flow starts in Part 1 as students consider what resources cells in the human body need. Students create a model describing the resources cells need and a **resource delivery** system to cells. Students revise the model as they gather information. Two resources needed by cells are **food (glucose) provided by the digestive system** (villi in small intestine to capillaries) and **oxygen provided by the respiratory system** (aveoli in lungs to capillaries). The **circulatory system uses blood vessels to transport** these resources to the cells.

In Part 2, students find out what happens once the glucose and oxygen get to the cells. They learn that in a cell, a chemical reaction takes place and the result is **usable energy for the cell (ATP) and waste products, water, and carbon dioxide**. This process to convert the energy in food into usable energy for the cell is called **aerobic cellular respiration**. Students learn about how organ systems interact to **remove the waste products**. The blood of the **circulatory system transports the waste products** to the lungs of the respiratory system and the **kidneys of the excretory system**. The process of aerobic cellular respiration occurs in most organisms, including plants. Plants are the **primary producers of glucose through the process of photosynthesis**.

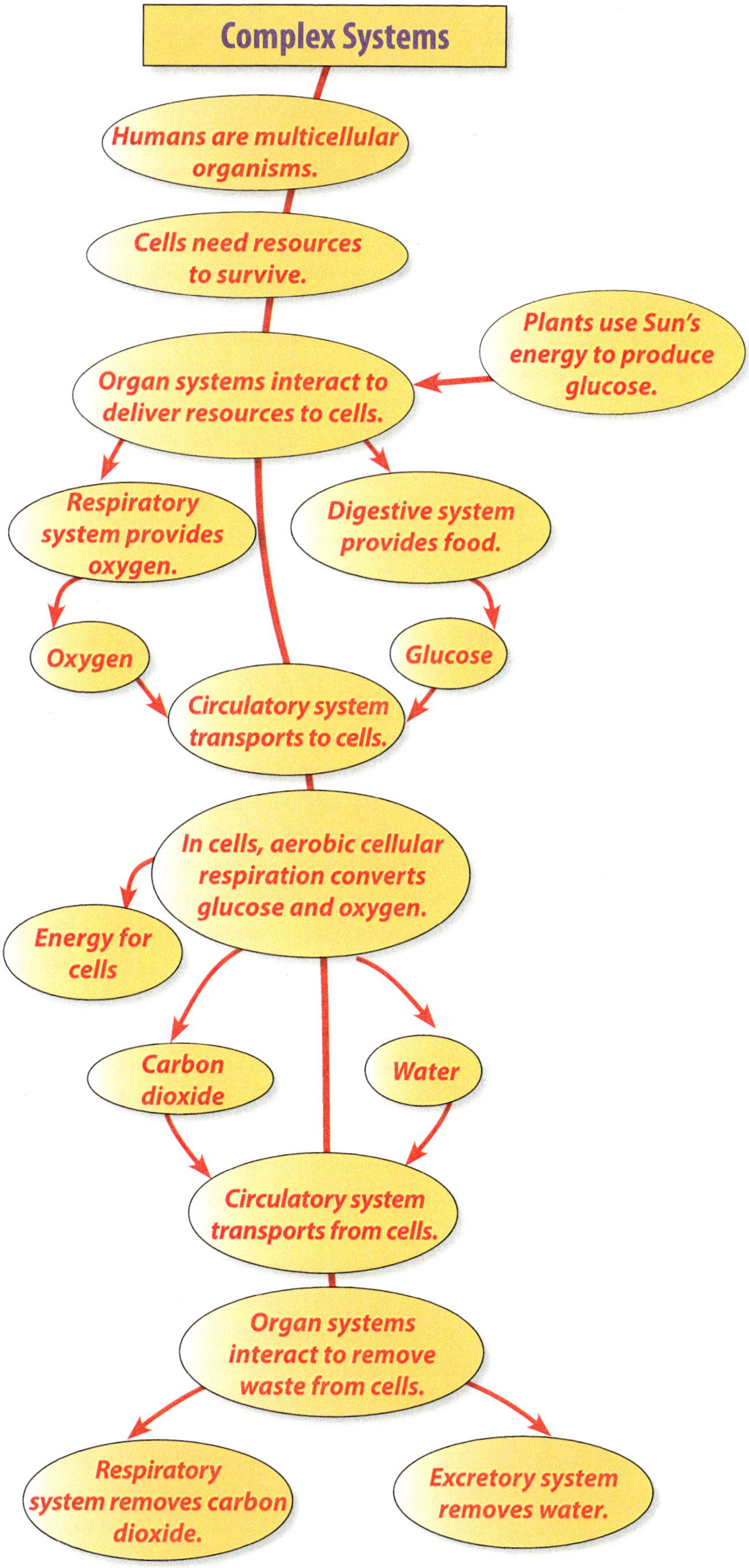

Human Systems Interactions Course—FOSS Next Generation

109

INVESTIGATION 2 — Supporting Cells

MATERIALS for
Part 1: Food and Oxygen

Provided equipment
For the class
1. Stopwatch

Teacher-supplied items
For each group
- Computers with Internet access
- Colored marking pens
- 4 Self-stick notes
- 3 Pieces of chart paper or ledger paper (28 × 43 cm; 11 × 17 in)

For the class
1. Bottle of water with nutrition label
1. Nutrition label

FOSSweb resources
For each student
1. Notebook sheet 4, *Organ-Systems Video Questions*
1. Notebook sheet 5, *Response Sheet—Investigation 2*

For the class
- Video, *Digestive and Excretory Systems*
- Video, *Circulatory and Respiratory Systems*
- Online activity: "Human Cardiovascular System" (in "Levels of Complexity")

For the teacher
1. *Embedded Assessment Notes*
- Teaching slides, 2.1

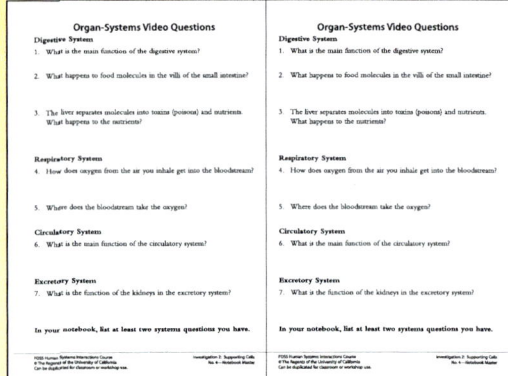

No. 4—Notebook Master

No. 5—Notebook Master

110　　　　　　　　　　　　　　　　　　　　　　　　Full Option Science System

Part 1: Food and Oxygen

GETTING READY for
Part 1: *Food and Oxygen*

Quick Start

Schedule	• Computers with Internet access, one for each group • 3 sessions active investigation
Preview	• Preview the FOSSweb Resources by Investigation for this part (such as printable masters, teaching slides, and multimedia) • Preview the videos: *Digestive and Excretory Systems*, Step 12 *Circulatory and Respiratory Systems*, Step 13 • Preview online activity: "Human Cardiovascular System," Step 15 • Preview response sheet and consider it for homework, Steps 18 and 21
Print or Copy	**For each student** • Notebook sheets 4, 5 **For the teacher** • *Embedded Assessment Notes*
Prepare Material	• Plan for physical activity **A** • Obtain nutrition labels and bottle of water. **B**
Plan for Assessment	• Review Step 18, "What to Look For" in response sheet **C**

▶ **NOTE**
Schedule computers for each group for Parts 1 and 2.

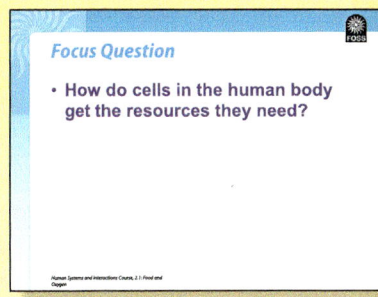

▶ **NOTE**
Preview the teaching slides on FOSSweb for this part.

Human Systems Interactions Course—FOSS Next Generation

INVESTIGATION 2 – Supporting Cells

Preparation Details

A Plan for physical activity

In Step 6, students will engage in some physical activity, running in place. Make sure that there is room around desks or tables so that students can run in place safely.

Consider options for students who are not physically able to run in place. Raising and lowering arms is a good alternative.

B Obtain nutrition labels and bottle of water

In Step 10, students will look at a nutrition label in their group to confirm that food has calories. You will need one nutrition label per group to use throughout the day. You can use labels from any food or drink package.

You will also need one nutrition label from a bottle of water to confirm that water has zero calories. You need only one bottle to use throughout the day in all your classes.

C Plan assessment: response sheet

Use notebook sheet 5, *Response Sheet—Investigation 2*, for a closer look at students' understanding of the interactions among the digestive, respiratory, and the circulatory systems. The response sheet can be completed in class or given as homework.

Preview What to Look For in Step 18 of Guiding the Investigation. Collect a sample of notebooks from each class, and after class, make notes about student progress on a copy of *Embedded Assessment Notes*.

Part 1: Food and Oxygen

GUIDING *the Investigation*
Part 1: *Food and Oxygen*

SESSION 1

Students will . . .
- Review organ-system interactions (Step 1)
- Write ideas about how cells in the body get resources (Step 2)
- Describe a resource delivery model for muscle cells (Step 3)
- Acquire information about heart and breathing rates during rest and exercise to understand the need for energy (Steps 4–9)
- Gather information from nutrition labels about energy (Step 10)

SESSION 2

Students will . . .
- Discuss needs of cells (Step 11)
- Gather and summarize information about human organ systems (digestive, circulatory, respiratory) from videos (Steps 12–14)
- Gather more information using online activity "Human Cardiovascular System" (Step 15)

SESSION 3

Students will . . .
- Revisit group resource delivery models (Step 16)
- Contribute to class presentation of a group model (Step 17)
- Demonstrate understanding on response sheet as assessment (Step 18)
- Review vocabulary and answer the focus question (Steps 19, 20)
- Review response sheet as homework (Step 21)
- Share notebook entries, discuss, and revise (Step 22)

FOCUS QUESTION
How do cells in the human body get the resources they need?

SESSION 1 45–50 minutes

1. **Review organ-system interactions**
   Give students a few minutes to review their last notebook entries with a partner. Ask them to discuss how thinking about the body in terms of systems and subsystems is useful for understanding how the human body functions.

CROSSCUTTING CONCEPTS

Systems and system models

Human Systems Interactions Course—FOSS Next Generation

113

INVESTIGATION 2 – *Supporting Cells*

SCIENCE AND ENGINEERING PRACTICES

Developing and using models

EL NOTE

Ask students to feel the muscle in their own calf or thigh.

TEACHING NOTE

"Resources" can remain general at this time. Students will consider the specific resources needed for aerobic cellular respiration (oxygen and food) in Steps 8–10.

Some students, especially those who researched the circulatory system, may already have a beginning handle on the "how." Do not confirm or dismiss any responses at this time. Simply collect the chart papers for future work.

TEACHING NOTE

Demonstrate both methods of finding a pulse and allow students to choose the method they are most comfortable with.

2. **Focus question: How do cells in the human body get the resources they need?**
Emphasize that it is important that cells get the resources they need to survive and function properly. Project or write the focus question on the board; have students write it in their notebooks.

➤ *How do cells in the human body get the resources they need?*

Give students a moment to consider the focus question and jot down their initial ideas. They should leave the rest of the page to return to later.

3. **Describe a resource delivery model**
Tell students that in order to come to a definitive conclusion, they will focus on particular cells, the cells in a leg muscle. Write these questions on the board. Have students discuss them in their groups.

➤ *What resources does each muscle cell in your leg need?*

➤ *How does each muscle cell get the resources it needs?*

Tell students,

To help us answer these questions, let's start with a model.

Explain that each group will first list the resources that they think muscle cells need and then create a diagram to explain how the muscle cells in a leg muscle get those resources. They should include a description of their model for a system that delivers resources to the cells. Explain that this is a first attempt and that they will return to the model later to add new information. Do not offer direction, but cruise the groups to see what they are thinking. This should move fairly quickly. Give students no more than 5–10 minutes. Provide each group with a large piece of paper or chart paper and marking pens.

When groups are done, ask them to put their period and group number or names on their models, and collect them. Groups will revisit their models later in this part.

4. **Take pulse rate**
Tell students that in order to consider these models further, they will have to acquire some information.

Show students two methods to find a pulse. 1) Show them how to take their wrist pulse. Explain that the best way to do this is to put the hand that is taking the pulse behind the other hand, and then wrap the fingers around the side of the wrist under the thumb. They should feel a throb under their fingertips, not the thumb (the thumb has a pulse of its own). Refer to the illustrations on the next page.

114

Full Option Science System

Part 1: Food and Oxygen

2) Students can also feel for their carotid artery. They can use their index and middle finger to find the pulse on either side of their neck in the groove between the windpipe and large neck muscle.

After students find their pulse, pose these questions for them to discuss with a partner.

▶ *Why do we feel a pulse?* [It has something to do with the heart beating.]

▶ *Which organ system do the heart and blood vessels belong to?* [Circulatory system.]

▶ *What is the function of the circulatory system?* [To get blood to all parts of the body, to all cells in the body.]

Confirm that the pulse students are feeling is related to the heart's pumping blood through the blood vessels. They can feel how slowly or quickly their heart is beating by counting their pulse. Ask students to consider how they could determine their pulse rate. Give them a minute to discuss a procedure to find out how many times their heart beats in a minute.

Have groups share their ideas. Then suggest that they will count how many times their heart beats in 15 seconds and then multiply by four. Give students a chance to find their pulse. Ask them to look at you once they are ready. Tell them to count silently.

When students are ready, start timing. Call "start" and then "stop" when 15 seconds are up. Tell students to record the number they counted and multiply by four to obtain their pulse rate. They should record their pulse rate in their notebook next to the notation: *Resting heart rate*. Alternatively, you can have students prepare a table such as the one in the sidebar. Give them time to record their results and share pulse rates with each other.

5. **Describe breathing**
Tell students that they also need to consider their breathing in order to think about the cells in the muscular system. Ask them to close their eyes and pay attention to their breathing for a few seconds without changing their breathing rate. After a time, ask,

▶ *What words would you use to describe your breathing?* [Air coming in and air leaving, feeling the chest rise and fall, relaxed, even, slow, regular.]

▶ *What parts of your body appeared to be involved in breathing?* [Lungs, nose, mouth, chest, abdomen (may say stomach), diaphragm.]

Ask students to record their observations in their notebooks.

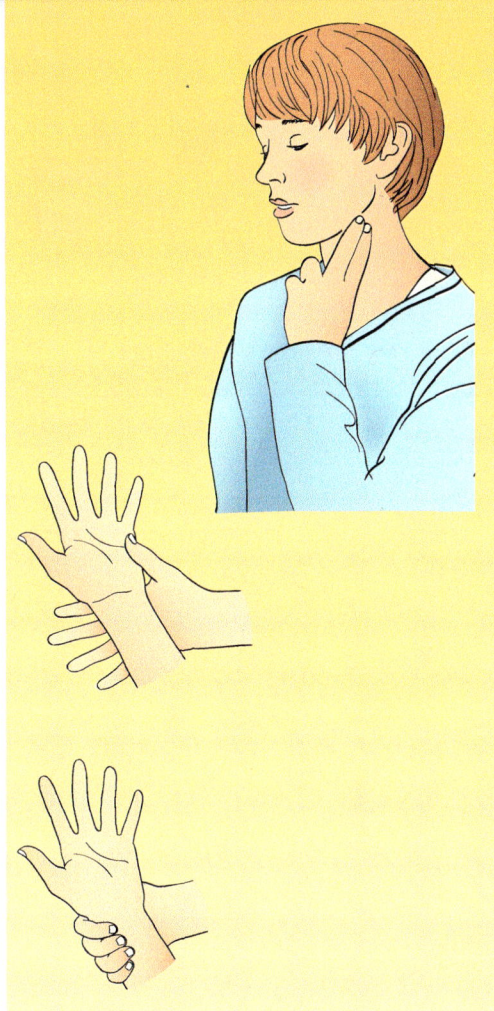

E L N O T E

Provide a recording chart as a way to organize their data.

Resting heart rate	Breathing description before exercise
Heart rate after exercise	Breathing description during and after exercise

Human Systems Interactions Course—FOSS Next Generation

INVESTIGATION 2 – Supporting Cells

> **TEACHING NOTE**
>
> Consider students who might not be able to run in place. They can do any activity that increases their pulse. One suggestion is to raise and lower their arms.

6. Use muscles

Tell students that they will use their muscles and observe what happens to their pulse rate and breathing. Ask them to clear the area around them. The challenge is to run quickly in place for an extended period of time that you determine (lifting the knees high is a very effective alternative). Students will record their pulse rate and breathing observations afterward. Stop the activity when you notice that students are starting to struggle (after a minute or so).

7. Take pulse and describe breathing

Immediately, have students count their pulse for 15 seconds, multiply by four, and record their pulse rate in their notebooks.

Tell students to record how their breathing changed during and after exercise.

Return to the heart rate. Ask,

➤ *What happened to your pulse rate after exercise?* [It increased.]

Confirm that pulse rates increased. Do not yet ask students to infer what an increased pulse rate might mean.

8. Confirm need for oxygen

Consider students' breathing observations. Ask,

➤ *What happened to your breathing during and after exercise?* [I had to breathe harder.]

➤ *Why did you start to breathe harder?* [Because I was moving a lot. I needed more air. I couldn't get enough air.]

Confirm that humans need oxygen from the air, and they need more oxygen while exercising, so they inhale more air.

Have students discuss with a partner why they think their pulse rate increased.

➤ *Why did your pulse rate increase?* [The heart had to pump faster to get oxygen to the muscles.]

9. Confirm need for energy

Ask students to think about running during physical education class. Ask,

➤ *Why might your bodies get tired or your muscles start hurting during exercise?* [They might be running out of energy.]

➤ *How does the body get energy?* [Accept all answers.]

> **TEACHING NOTE**
>
> It is a common misconception that humans can get energy from water, resting, or caffeine. Caffeine is a stimulant but provides no calories. You will cross all of these off the list at the end of Step 10.

116   Full Option Science System

Part 1: Food and Oxygen

Write student responses on the board. You will return to the responses at the end of Step 10.

Tell students to focus on food for a moment. Ask, *on slide*

➤ *What is the importance of food for our bodies?* [Food gives us nutrition and energy.]

Confirm the body's need for energy.

10. Discuss food as a source of energy

Distribute one nutrition label to each group. Ask groups to look at the label they have and identify how much energy is in that food. If students do not refer to calories, tell them,

Food is a source of energy. Food is the only source of energy for animals, including humans. The unit we use to measure food energy is the **calorie**. *The more calories a food has, the more energy it has.*

Look for the calories on your food labels.

Ask for a few groups to report the number of calories per serving their food has. Emphasize that there is no other way for them to get energy. Have students write in their notebooks, "Energy comes from food. Energy is measured in calories."

Hold up the bottle of water. Ask,

➤ *Does water have energy?*

Have a student look at the nutrition label on the bottle of water and tell the class how many calories it has (zero). Ask,

➤ *Does water have energy?* [No.]

Confirm that water has no calories and is therefore *not* a source of energy, even if it is essential for life. Cross water off the list.

➤ *Does sleep provide your body with energy?* [No.]

Tell students that sleep helps your body function better, but does not provide energy. Cross sleep and any other processes or substances except for food off the list. To get energy, one must eat food or drink beverages with calories.

EL NOTE

Add "calorie" to the word wall. Students may want to make a word map for "energy." See the example below.

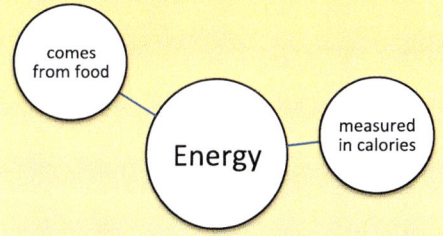

TEACHING NOTE

Cells depend on water for chemical reactions to occur, and the body uses water to regulate temperature and other bodily functions.

SCIENCE AND ENGINEERING PRACTICES

Constructing explanations

Human Systems Interactions Course—FOSS Next Generation

INVESTIGATION 2 – *Supporting Cells*

SESSION 2   45–50 minutes

11. Discuss cells' needs

Ask students to think about the running activity. Ask,

➤ *What two resources did we identify that the body needs?* [Oxygen and food.]

Give groups a minute to consider the following questions before calling for responses.

➤ *Do muscle cells in the leg need oxygen and food? What is your reasoning?* [Yes. If the body needs those resources, every cell needs them.]

➤ *What about cells that make up your lungs, or your brain, or your stomach? Do they all need oxygen and food as well?* [Yes.]

12. View video: *Digestive and Excretory Systems*

Remind students that they are trying to figure out how each muscle cell gets the resources (food and oxygen) it needs. Tell them that they will first think more about how food gets to the muscle cells.

Distribute notebook sheet 4, *Organ-Systems Video Questions.* (Students will answer items 1–6 in this part of the investigation and complete the sheet in Part 2.) Preview items 1–3 with students. Play the three *Digestive and Excretory Systems* video chapters in "Resources by Investigation." You may need to pause at the appropriate spots for students to record their responses. Here are the chapters.

- Chapters 2 and 3: "Mouth" and "Esophagus" and "Stomach" (duration 4.5 minutes)

- Chapter 5: "Small Intestine" (2 minutes)

Confirm that the digestive system breaks down food into molecules that cells can use. The molecules pass into the bloodstream from the small intestine (the jejunum section of the small intestine in the video) to the liver, where they are separated into toxins and nutrients. The nutrient molecules go back into the bloodstream and travel to the cells (via the heart).

Pause the video at the diagrams to allow students to make quick sketches in their notebook to label and show the pathway of food to molecules to delivery to cells. Students can also use the section on the digestive system in the *FOSS Science Resources* book to add more information to their diagrams.

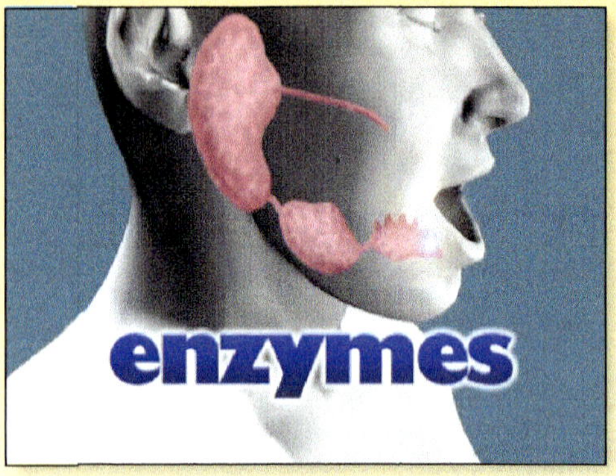

ELA CONNECTION

This suggested strategy addresses the Common Core State Standards for ELA in literacy.

WHST 8: Gather relevant information from multiple print and digital sources.

Part 1: Food and Oxygen

13. **View video:** *Circulatory and Respiratory Systems*

    Now students need to think about the oxygen that cells need. Direct their attention to items 4 and 5 on the notebook sheet.

    Play chapter 3, "Respiratory System" (duration 2 minutes) from the *Circulatory and Respiratory Systems* video.

    Confirm that oxygen transfers into the bloodstream from the lungs. Air sacs called **alveoli** are surrounded by **capillaries**, which are the tiny blood vessels that start the oxygen on its way to the cells.

    Give students a few minutes to answer the questions on their notebook sheet. Pause the video at the diagrams to allow students to make quick sketches in their notebooks to label and show them how oxygen is transferred to the bloodstream. Students can also use the section on the respiratory system in the *FOSS Science Resources* book to add more information to their diagrams.

    Preview item 6 on the notebook sheet.

    Play chapter 5, "Heart" (duration 3 minutes). It describes how the heart pumps blood that is rich in oxygen and nutrients to the body.

    Follow the same procedure, giving students a few minutes to answer the question on their notebook sheets. Pause the video at the diagrams to allow students to make quick sketches in their notebooks to label and show how blood travels into, through, and out of the heart. Students can also use the section on the circulatory system in the *FOSS Science Resources* book to add more information to their diagrams.

    Have students write in their notebooks questions they have.

SCIENCE AND ENGINEERING PRACTICES

Obtaining, evaluating, and communicating information

14. **Summarize the videos**

    Give students a few minutes in their groups to talk about questions 1–6 on the notebook sheet before summarizing as a class.

    ➤ *What organ systems are involved in getting food to every cell in the body?* [Digestive system and circulatory system.]

    ➤ *How does that work?* [The digestive system breaks down the food we eat into nutrient molecules, which transfer to the blood through the small intestine and travel to the cells in the body.]

    ➤ *What organ systems are involved in getting oxygen to every cell in the body?* [Respiratory system and circulatory system.]

    ➤ *How does that work?* [Oxygen comes into our bodies through the nose and mouth and goes to the lungs. Oxygen transfers to blood in capillaries at the alveoli and travels to cells in the body.]

TEACHING NOTE

Note that the last question will be answered in Part 2. Let students know that as well.

CROSSCUTTING CONCEPTS

Systems and system models

Human Systems Interactions Course—FOSS Next Generation

INVESTIGATION 2 – Supporting Cells

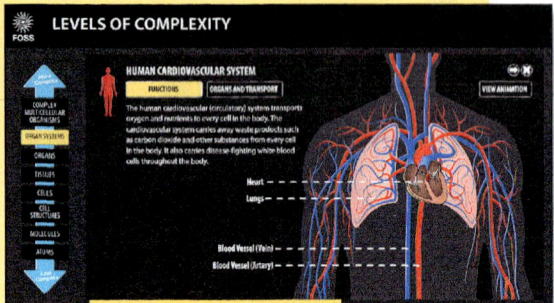

TEACHING NOTE

Remind students that the cardiovascular system is part of the circulatory system.

SCIENCE AND ENGINEERING PRACTICES

Constructing explanations

Obtaining, evaluating, and communicating information

15. View online activity: "Human Cardiovascular System"

Tell students that you have one final way for them to think about how cells get the oxygen and food they need. If possible, make a computer available to each group.

Access the "Human Cardiovascular System" activity through "Resources by Investigation." Click the human figure to enter the activity. Remind students that another name for this part of the circulatory system is the cardiovascular system.

Have students toggle back and forth between "Functions" and "Organs and Transport." In "Organs and Transport," point out the white arrows, which indicate the flow of blood throughout the body. Ask a student to read the paragraph that describes the flow of blood. Stop the student immediately after the first sentence and ask about the term "nutrient-rich blood."

Pose these questions for students to discuss.

➤ *Where does the blood pick up nutrients?* [Small intestine.]

➤ *Where do those nutrients come from?* [The food we eat.]

➤ *Where does the blood go to after it has nutrients in it?* [The lungs.]

Have the student finish reading the rest of the paragraph. Ask,

➤ *After the blood has acquired nutrients and oxygen, what function does it serve?* [It transports resources to the cells in order for the cells to produce energy.]

Have students click the View Animation button. The animation shows various resources being transported by the capillaries to cells in the body. Focus on "Needs Energy" and "Exchanges Gases." Note that the pink cells represent the walls of the capillaries and the yellow cells are generalized tissue cells.

Tell students that these energy and gas exchanges are happening at the same time.

Part 1: Food and Oxygen

SESSION 3   45–50 minutes

16. Revisit group resource delivery models

Tell students that now they will take a second look at their models for muscle cells to get the resources they need. Hand back the original resource delivery models that groups created. Distribute two new pieces of chart paper or large paper to each group.

Tell students that this time they will work in pairs on the revision. Tell them that they will present their final model to the rest of the class. When groups share their work, they can decide which pathway they would like to present to the class.

Ask students to write "food" and "oxygen" at the top of their papers. These are the only two resources that they should consider at this time. They should model the process that delivers each substance to cells in a leg muscle.

Write on the board the vocabulary they may want to include (see the sidebar for examples). Let students decide how to prepare this representation. (If groups are stuck, you might show them how they can sketch a body outline to start.) Give groups about 15 minutes to prepare their models, incorporating the information they have learned from the videos, discussion, and online activity.

NOTE: Watch for the misconception that oxygen and food molecules travel in different blood vessels.

17. Present group models

Call on each group to present one of their models and use it to explain to the rest of the class how nutrients or oxygen enter and travel through the body to the leg muscle cell. Encourage other groups to ask questions for clarification and elaboration.

18. Assess progress: response sheet

Distribute notebook sheet 5, *Response Sheet—Investigation 2*, to each student (but don't have them tape or glue them into their notebooks at this time). Assign the response sheet as homework, or give students time to respond to the prompt in class. Collect the sheets and use them to consider students' thinking about how the digestive and respiratory systems work with the circulatory system.

What to Look For

- Students recognize that while there are "air tubes" to bring oxygen into the body (mouth and nose to through trachea to the lungs) and "food tubes" to get food into the body (mouth, esophagus to stomach), that's where the process of simple tubes ends.

---

alveoli
bloodstream
capillary
cell in muscle
digestive system
lungs
respiratory system
small intestine
villi

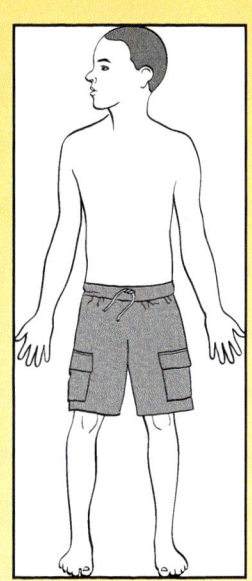

EL NOTE

For students who need scaffolding, provide a sentence frame such as
"What does _____ represent?"
"Why did you include _____?"
"Did you consider _____?"
"Can you explain _____?"

TEACHING NOTE

Students can use the back of the sheet to complete their responses. After you review and return sheets to students, they can attach one edge of the sheet to the notebook so the back is still accessible.

Human Systems Interactions Course—FOSS Next Generation

INVESTIGATION 2 – Supporting Cells

SCIENCE AND ENGINEERING PRACTICES

Constructing explanations

Obtaining, evaluating, and communicating information

TEACHING NOTE

Refer to the Science-Centered Language Development in Middle School and the Science Notebooks in Middle School chapters in Teacher Resources for additional information on reviewing vocabulary and next-step strategies.

alveolus
calorie
capillary

ELA CONNECTION

These suggested strategies address the Common Core State Standards for ELA for literacy.

SL 1: Engage in collaborative discussion.

WHST 5: Develop and strengthen writing.

L 6: Acquire and use academic and domain-specific words.

- *Students describe a model of how food is broken down and moves through humans: food enters through the mouth when digestion begins; transfers to stomach, then small intestine; in intestine, food is broken down into nutrient molecules, which are absorbed into bloodstream through capillaries.*
- *Oxygen enters the body through the mouth and nose; travels to lungs and small sacs (alveoli); absorbed into bloodstream through capillaries.*
- *Blood and blood vessels carry oxygen and food molecules to capillaries that provide these resources to cells.*

Plan to spend 15 minutes outside of class to review a selected sample of student responses, using *Embedded Assessment Notes* to record your observations. After your review, return the notebook sheets to students to be taped or glued into their science notebooks Refer to the Assessment chapter for next-step strategies.

If a next-step strategy is needed, try having students critique anonymous student work. Show a student response that is incomplete/and or has errors. Let students discuss what is good about the response and how it could be better. Then let students revise their own responses.

19. Review vocabulary

Give students a few moments to review the vocabulary developed in this part. This is a good time to update their vocabulary indexes and tables of contents if they haven't already done so.

20. Answer the focus question

Ask students to summarize their findings and respond to the focus question.

➤ *How do cells in the human body get the resources they need?*

21. Extend the investigation with homework

Send home notebook sheet 5, *Response Sheet—Investigation 2*, as homework if students did not have time to complete or revise it in class.

WRAP-UP/WARM-UP

22. Share notebook entries

At the end of this session or the beginning of the next, have students exchange notebooks with a partner and conduct a peer review of the answer to the focus question. Each student writes on a self-stick note a question or prompt to help his/her partner strengthen their writing. Allow students a few minutes to revise their response based on the feedback from their partner.

122   Full Option Science System

Part 2: Aerobic Cellular Respiration

MATERIALS for
Part 2: Aerobic Cellular Respiration

Provided equipment
For each student
- 1 *FOSS Science Resources: Human Systems Interactions*
- • "Aerobic Cellular Respiration"

For each group
- 1 Transparent tape roll

Teacher-supplied items
For each student
- 1 Piece of white paper, 22 × 28 cm (8.5 × 11 in)
- 1 Strip of tiles (made from teacher master B)

For each group
- • Colored pencils (optional)
- 2 Scissors
- • Computers with Internet access
- • Self-stick notes (optional, See Step 14)

For the class
- 1 Document camera

FOSSweb resources
For each student
- 1 Notebook sheet 6, *To and From the Cells*
- 1 Teacher master A, *Aerobic Cellular Respiration*
- 1 *Investigations 1–2 I-Check*

For the class
- • Video, *Digestive and Excretory Systems*

For the teacher
- • Teacher master B, *Aerobic Cellular Respiration Tiles*
- • Performance Assessment Checklist
- • Assessment Record
- • Teaching slides, 2.2

FOSS Science Resources

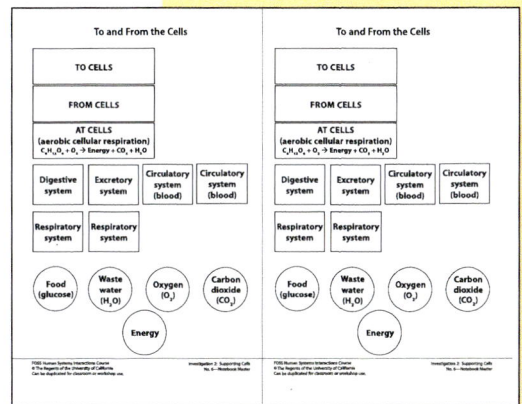

No. 6—Notebook Master

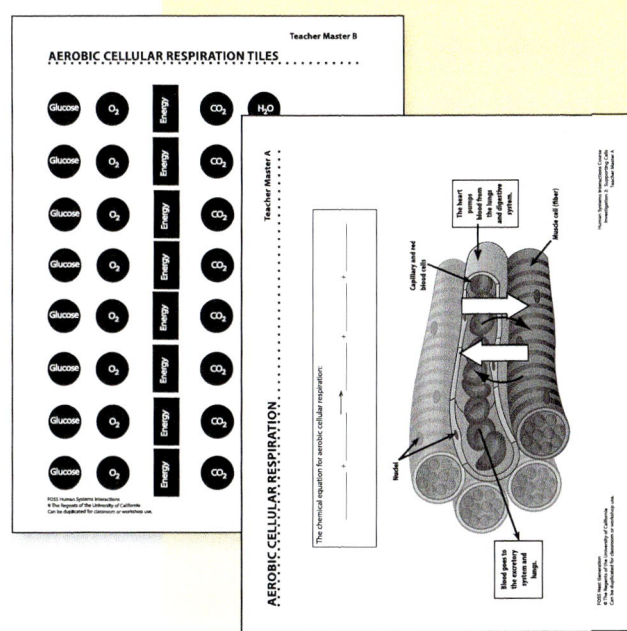

Teacher masters A and B

Human Systems Interactions Course—FOSS Next Generation

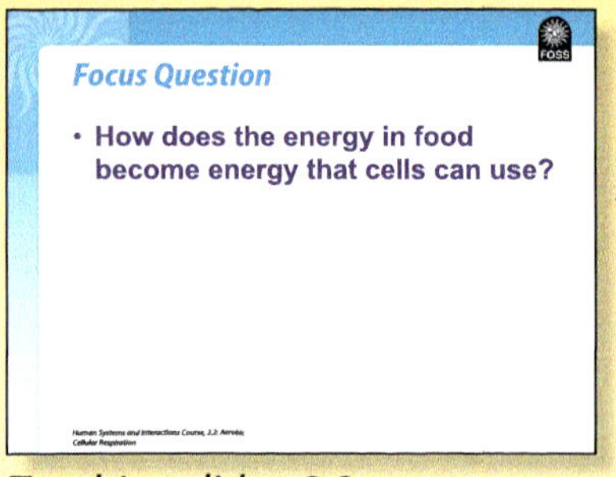

INVESTIGATION 2 – Supporting Cells

GETTING READY for
Part 2: Aerobic Cellular Respiration

Quick Start

Schedule	• Computers with Internet access, one for each group • 2 sessions active investigation • 2 sessions assessment
Preview	• Preview the FOSSweb Resources by Investigation for this part (such as printable masters, teaching slides, and multimedia) • Preview video: *Digestive and Excretory Systems*, Step 10 • Preview the reading: "Aerobic Cellular Respiration," Step 13
Print or Copy	**For each student** • Notebook sheet 6 • Teacher master A • *Investigations 1–2 I-Check* or schedule it on FOSSmap **For the teacher** • Teacher master B • *Performance Assessment Checklist* • *Assessment Record* **C**
Prepare Material	• Prepare cellular-respiration tiles **A** • Practice tiles activity **B**
Plan for Assessment	• Review Step 13, "What to Look For" in the performance assessment • Plan for benchmark assessment, Step 21, *Investigations 1–2 I-Check* **C**

Teaching slides, 2.2

Focus Question
• How does the energy in food become energy that cells can use?

Part 2: Aerobic Cellular Respiration

Preparation Details

A Prepare cellular-respiration tiles

Each copy of teacher master B, *Aerobic Cellular Respiration Tiles,* has enough tile strips for eight students. Make enough copies so that each student has one strip. Cut strips before distributing them.

Each student will use a copy of teacher master A, *Aerobic Cellular Respiration,* with the tiles. Make enough copies for each student.

B Practice tiles activity

Practice the modeling process in Guiding the Investigation, Steps 4–9, so that you can lead the class.

C Plan benchmark assessment: I-Check

Plan the process for administering and discussing the results of the *Investigations 1–2 I-Check.* The I-Check serves as a checkpoint for student learning. Refer to the Assessment Chapter for details.

If students will be taking the I-Check online in FOSSmap, open the assessment before class on the day students take the I-Check.

At least one day before taking the I-Check, allow time for students to review their notebook entries. When taking the I-Check, students should not use their notebooks, but the notebooks are a good tool to use when students later reflect on their answers.

To track achievement (a summative use of the I-Checks), use the coding guides in the Assessment chapter to code the items, or review the FOSSmap reports that automatically code most items (code open response questions before running the online reports).

I-Checks can also be used for formative assessment. Research has shown that students learn more when they take part in evaluating their own responses. When students check their own understanding, you are creating a class culture of assessment as a tool in the service of learning. Here's how to do this with paper I-checks.

- Have students complete an I-Check unassisted.
- Code I-Check item, but do not make any marks on students' written responses. Record codes on *Assessment Record* or a grade book. Note important points about the items to review with students.
- Return I-Checks to students. Use self-assessment strategies as described in the Assessment chapter to facilitate reflection and clarify student thinking.

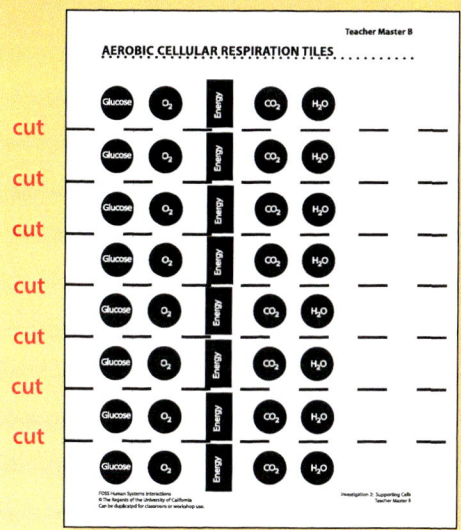

Teacher master B

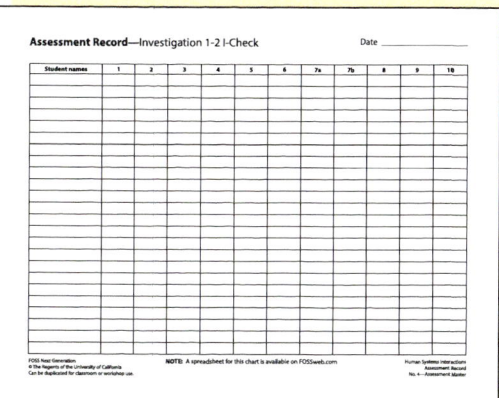

Assessment Record

Human Systems Interactions Course—FOSS Next Generation

125

INVESTIGATION 2 – Supporting Cells

FOCUS QUESTION

How does the energy in food become energy that cells can use?

GUIDING the Investigation
Part 2: Aerobic Cellular Respiration

SESSION 1

Students will . . .
- Discuss how energy in food becomes energy for cells (Steps 1, 2)
- Use a model to understand aerobic cellular respiration (Steps 3–8)
- Identify the products of aerobic cellular respiration (Steps 9,10)
- Gather information about the excretory system from video (Step 11)
- Summarize the process of aerobic cellular respiration through a model (Steps 12, 13)

SESSION 2

Students will . . .
- Gather and communicate information by reading and discussing "Aerobic Cellular Respiration" (Steps 14–16)
- Review vocabulary and answer the focus question (Steps 17, 18)
- Review notebook entries and generate a list of big ideas (Step 19)

SESSIONS 3–4

Students will . . .
- Demonstrate understandings by responding to *Investigations 1–2 I-Check* (Step 20)
- Review I-Check items through next-step strategies (Step 21)

SESSION 1   45–50 minutes

1. **Move to the cells**
   Hold up one of the student pathway models from the previous part, and pose these questions.

   ➤ *Food and oxygen molecules have been delivered to the cells. What do the muscle cells do with them?* [Students may not know.]

   ➤ *What did we say food is a source of?* [Energy.]

   Tell students that the energy in food molecules needs to be transformed to a form that cells can use.

Part 2: Aerobic Cellular Respiration

2. **Focus question: How does the energy in food become energy that cells can use?**

   Pose the focus question, write or project it on the board, and have students write it in their notebooks.

   ➤ *How does the energy in food become energy that cells can use?*

   Students do not need to answer the question at this time. They will answer the focus question at the end of this part. They should leave space under the question to answer it later.

3. **Introduce** *aerobic cellular respiration*

   Tell students,

   The food we eat is turned into usable energy using a chemical process called **aerobic cellular respiration**.

   Write the words on the board. Ask students to work in their groups to quickly define each word or list words that might be associated with each word. When most groups are finished, ask,

   ➤ *What do you think of when you think of the word, "aerobic?"* [Gym; air.]

   aerobic: requires oxygen
   cellular: happens in cells
   respiration: how cells get usable energy from food molecules

   Tell students to write the word in their notebooks on a new page. You might want to have them use a Frayer model to help them develop this concept. Students can add to this as you move through Step 3. See the completed example on page 139.

   Tell them that any aerobic activity requires oxygen. They should write that down. Ask,

   ➤ *What about cellular?* [In cells.]

   ➤ *Where do you think aerobic cellular respiration occurs?* [In cells.]

   Tell students to write that down as well. Ask,

   ➤ *What comes to mind when you think of respiration?* [Respiratory system, breathing, oxygen.]

   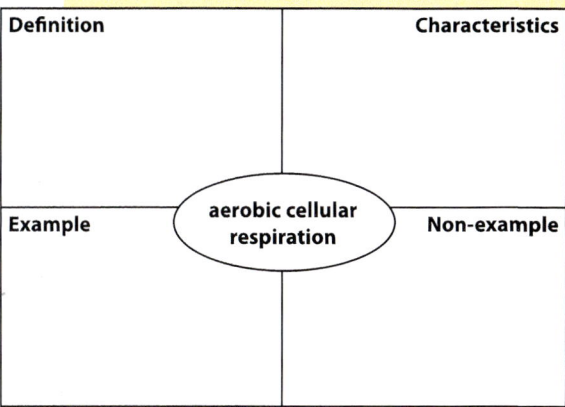

   Tell students that respiration indeed is connected to breathing. But cells themselves don't breathe. The words *cellular respiration* in the context of cells refers specifically to the process by which cells use oxygen to obtain usable energy from food molecules.

   Now students have a working definition of aerobic cellular respiration: the process that cells use to turn food molecules and oxygen into energy that they can use to grow, repair, move, and go about the business of living. Tell students to write the working definition in their science notebooks and record the phrase in their vocabulary indexes.

   ELA CONNECTION

   This suggested strategy addresses the Common Core State Standards for ELA for literacy.

   L 6: Acquire and use academic and domain-specific words.

   CROSSCUTTING CONCEPTS

   Scale, proportion, and quantity

Human Systems Interactions Course—FOSS Next Generation

INVESTIGATION 2 – *Supporting Cells*

4. **Introduce modeling aerobic cellular respiration**

    Distribute a copy of teacher master A, *Aerobic Cellular Respiration*, to each student. Project a copy of the teacher master using a document camera and orient students to the diagram. Identify the capillary and red blood cells, and point out that the long "tubes" surrounding the capillary are skeletal muscle cells, called muscle fibers. Ask,

    ➤ *How do oxygen and food get into the blood?* [Oxygen comes from the air and enters through the respiratory system; food is eaten and enters from the digestive system. The blood carries the oxygen and nutrients through the circulatory system to the cells.]

    Ask students to point to where blood from the lungs (respiratory system), digestive system, and heart (circulatory system) is coming from (right side of the diagram).

5. **Identify the reactants**

    Focus students' attention to the chemical equation at the top of teacher master A. Ask them to identify the reactants in this process.

    If students need help you can provide this information.

    *The food we eat is broken down into three kinds of nutrient molecules: carbohydrates, fats, and proteins. The main source of energy for your cells is a sugar molecule called **glucose**. Glucose comes from the food we eat. The glucose is transported to the cells by the blood. Oxygen is brought into the body by the respiratory system and transported to the cells by the blood.*

    Ask students to write the reactants in the chemical equation at the top of the sheet.

    glucose + oxygen ⟶ ____ + ____

6. **Identify the products**

    Ask,

    ➤ *What do you think the products of this reaction are? What do the oxygen and glucose change into?* [Students should know energy as that was part of the definition of aerobic cellular respiration. Explain that the other two products are carbon dioxide and water.]

    Ask students to write the products in the chemical equation at the top of the sheet.

    glucose + oxygen ⟶ carbon dioxide + water + **energy**

> **TEACHING NOTE**
>
> Energy is stored in the molecule adenosine triphosphate (ATP). It is a readily available energy source for many cellular metabolic processes. It is not necessary to identify this for students. They will encounter that molecule in high school biology.

Part 2: Aerobic Cellular Respiration

7. Begin the modeling

Tell students that you have a set of symbols for them to use to model how the energy in food becomes usable energy for cells. Ask Getters to get scissors for their groups and a strip of tiles for each person. Review the symbols on the tiles.

Help students get started by using your copy of the teacher master. Place the glucose tile and the oxygen tile on the right end of the capillary. Ask students to work with their partner to move the symbols through the illustration to model the process.

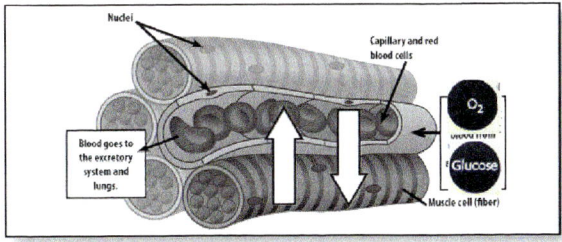

Give students enough time to design their model and move the tiles. Have them share their model with a partner. Circulate among the groups and provide support as necessary.

8. Review the model

When groups have completed their models, call for attention. Ask a pair of students to come up to the document camera to share how they modeled the process. You and the class can ask questions of the pair of students.

This is what a review demonstration should look like.

a. **Deliver resources to cell.** Blood carries glucose and oxygen to a capillary near the muscle cell. These resources pass through the capillary to the muscle cell. Record that movement of the molecules by writing the names on the arrow to the muscle cell.

b. **Aerobic cellular respiration in cell.** A chemical reaction takes place in the cells and products are produced. Those products are carbon dioxide, water, and energy. Those tiles should be on the muscle cell. The cells have the energy they need to function, to make the muscles move. The glucose and oxygen tiles are removed.

c. **Remove waste products from cell.** Move the water and carbon dioxide tiles from the muscle cell back to the capillary and move the molecules left along the capillary to the excretory system. Record the names of the waste products in the arrow going to the capillary.

SCIENCE AND ENGINEERING PRACTICES
Developing and using models

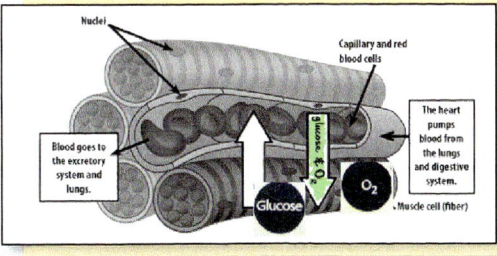

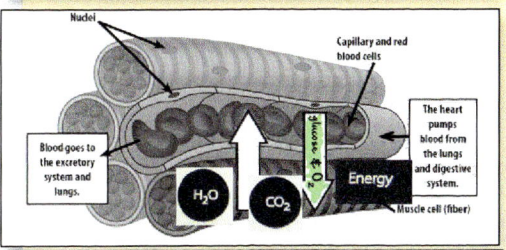

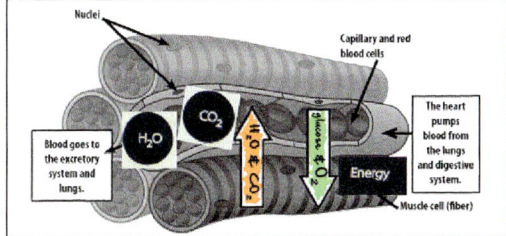

Human Systems Interactions Course—FOSS Next Generation

129

INVESTIGATION 2 – Supporting Cells

9. **Discuss products**

    Ask,

    ➤ *What are the waste products of aerobic cellular respiration?* [Carbon dioxide and water.]

    ➤ *How do our bodies get rid of carbon dioxide?* [Breathe it out.]

    ➤ *How does the carbon dioxide get to the lungs?* [Blood carries carbon dioxide to the lungs.]

    ➤ *How does your body get rid of extra water?* [Sweat, urine, water vapor in breath.]

    ➤ *What happens to the energy?* [It stays in the muscle cell and is used.]

    Be sure to emphasize this point. Energy does not just disappear into another part of the body. Cells use it for growth and repair.

    If necessary, provide this information.

    - Carbon dioxide gas and excess water are exchanged across the cellular membrane of each muscle cell into a capillary.
    - The blood carries the waste gas (including water vapor) to the alveoli in the lungs, where it is exhaled back into the atmosphere.
    - The blood carries the water through the excretory system. The excretory system (including the kidneys and liver) filters out waste products from the blood and removes them from the body.

10. **Work with the model**

    After the discussion, give students a few moments to work with a partner to manipulate the tiles to model the process and describe it to each other. They should write information about the process on the sheet so that it describes the model.

    When done, students should tape or glue all the tiles onto the chemical equation portion of the sheet and affix the sheet in their notebooks. Have them fold the page in half and attach it along one edge onto a left-hand page of their notebooks.

> **TEACHING NOTE**
>
> Note that metabolic wastes in urine are filtered out by the kidneys and the liver. This is why urine is not pure water. You cannot drink your urine for long before those metabolic wastes become more and more concentrated in the urine and cause your kidneys to fail.

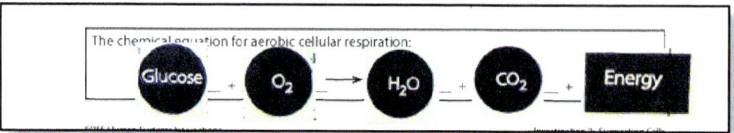

Part 2: Aerobic Cellular Respiration

11. **View video: *Digestive and Excretory Systems***

    Play chapter 7 from the video *Digestive and Excretory Systems* (duration 4.5 minutes, or you can stop after 2 minutes). This chapter describes how the excretory system eliminates bodily waste products.

    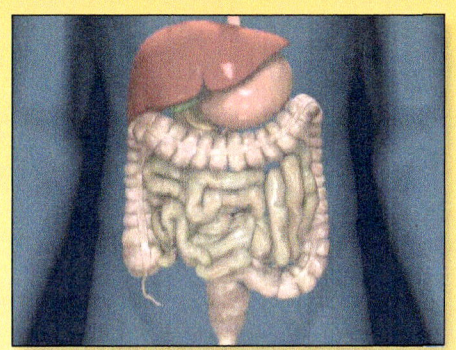

    Have students answer item 7 on notebook sheet 4, *Organ-Systems Video Questions*. When they are done, they should list any questions they still have. You might want to spend a few minutes addressing these questions before students take the I-Check.

12. **Summarize the cell-servicing process**

    Distribute a copy of notebook sheet 6, *To and From the Cells*, and one 8.5 × 11 inch piece of paper to each student. Tell students that their task is to summarize the process that provides food and oxygen to the cells and that eliminates the waste products of aerobic cellular respiration.

    Project a copy of the notebook sheet and give the following directions:

    On the notebook sheet, there are

    - *three location tiles (to cells, from cells, at cells)*
    - *six organ-system tiles*
    - *four substance tiles*
    - *one energy tile*

    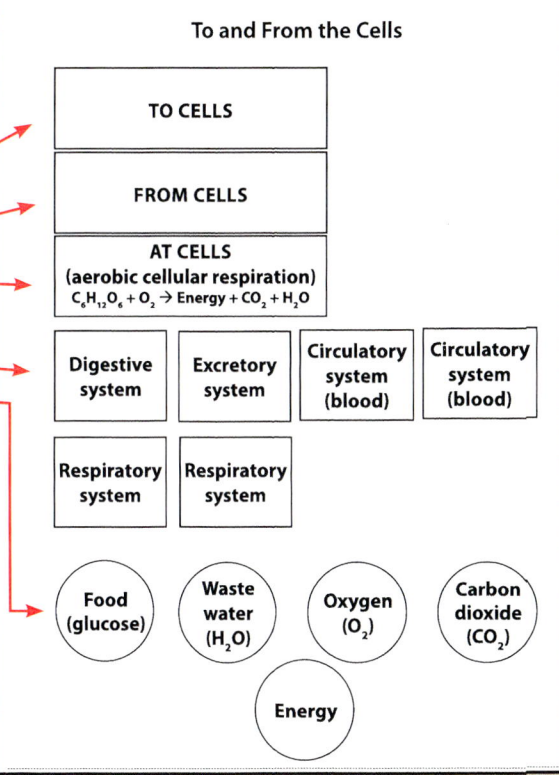

    Take care not to lose any of them. Cut out the tiles and arrange them to model how cells are serviced—resources are delivered, used by cells, and waste removed.

    When you are satisfied with your arrangement, transfer the tiles to an 8.5 × 11 piece of paper. Draw arrows where necessary to complete your model. Tape or glue the tiles down on the paper and include text to describe your cell-servicing model.

    As you cruise the room, have as many students as possible verbally relate their ideas to you before they tape down the tiles. Ask questions such as these.

    ➤ *Where in your model is energy released?*

    ➤ *What substances are necessary for aerobic cellular respiration?*

    ➤ *How does your model show where the substances come from?*

    ➤ *How does your model show how these get to the cells?*

CROSSCUTTING CONCEPTS

Systems and system models

Energy and matter

Human Systems Interactions Course—FOSS Next Generation

INVESTIGATION 2 – Supporting Cells

SCIENCE AND ENGINEERING PRACTICES

Developing and using models

DISCIPLINARY CORE IDEAS

LS1.A: Structure and function

LS1.C: Organization for matter and energy flow in organisms

CROSSCUTTING CONCEPTS

Systems and system models

Energy and matter

TEACHING NOTE

You will be amazed at the incredible variety of strategies that students employ to organize and communicate their understanding of how the organ systems work together to deliver the goods and remove waste.

We have provided just one example on the answer sheet displayed here. There are other ways to show this process.

13. **Assess progress: performance assessment**

During class, check students' ability to model how food, oxygen, and energy move through human systems.

What to Look For

- *Students draw and explain a model in which oxygen is brought in by the respiratory system and food (glucose) is brought in by the digestive system to each cell where energy is released. (Developing and using models; systems and system models.)*

- *The circulatory system collects the oxygen and food (glucose) and delivers them to the cells. (LS1.A: Structure and function.)*

- *There is a process within cells that uses oxygen and glucose to make energy available to cells (aerobic cellular respiration). (LS1.C: Organization for matter and energy flow in organisms; energy and matter.)*

- *The waste products of aerobic cellular respiration, carbon dioxide and water, are delivered by the circulatory system to the respiratory and excretory systems. Water exits the body by either the respiratory system (lungs) or the excretory system (urinary system); carbon dioxide exists the body largely through the respiratory system. (LS1.A: Structure and function.)*

Record your observations on the *Performance Assessment Checklist*. Plan next-step strategies as needed.

Students can fix their models into their science notebooks.

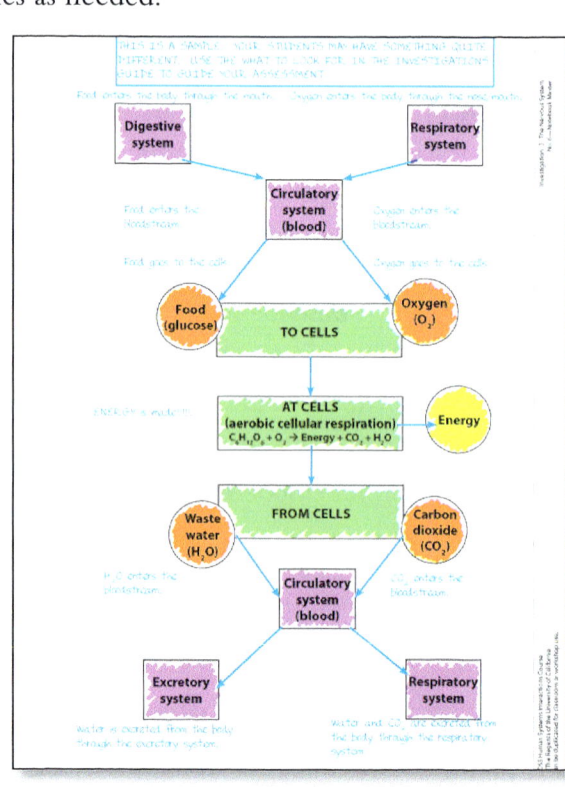

Part 2: Aerobic Cellular Respiration

SESSION 2 45–50 minutes

READING *in Science Resources*

14. Read "Aerobic Cellular Respiration"
Tell students that this article will give them more information about aerobic cellular respiration. Suggest they start by previewing the text features, including how the text is organized. Give them a few minutes to discuss the photographs, captions, and subtitles with a partner.

15. Use a reading comprehension strategy
If students are using the Frayer model, tell them to review what they have written so far and think about what additional information they need to help them understand aerobic cellular respiration. Give students self-stick notes to jot down their thoughts and questions as they read.

First read as individuals: Tell students to read the article independently, using the self-stick notes to mark any unknown words or phrases, questions they have, and important information they want to add to their notes. When students finish reading, have them talk over any difficulties they had with the text and share their questions.

Second read with whole class: Refer students back to the first page and ask for a volunteer to read the first sentence under the title. Continue reading by having volunteers read aloud. Follow the annotated guide on the next five pages, stopping at the designated letters to engage students in discussions about specific questions.

FOSS Science Resources

ELA CONNECTION
This suggested strategy addresses the Common Core State Standards for ELA in literacy.

L 6: Acquire and use academic and domain-specific words.

SCIENCE AND ENGINEERING PRACTICES
Obtaining, evaluating, and communicating information

TEACHING NOTE
For more information on scaffolding literacy in FOSS, see the Science-Centered Language Development in Middle School chapter in Teacher Resources.

Human Systems Interactions Course—FOSS Next Generation

133

INVESTIGATION 2 – Supporting Cells

Cite evidence: Ask students to think about the concept of aerobic cellular respiration in terms of matter and energy. The text says living things need energy, yet the photograph here shows a girl eating a sandwich, which is matter. Why is that? Reread this page to find evidence to support your ideas. [Food is made of molecules (matter). Energy is stored in the chemical bonds of those molecules and is released during chemical reactions.]

Analyze purpose in a text:

➤ What substances are needed to make usable energy in a cell? [Oxygen and glucose.]

Determine the meaning of symbols: Look at the chemical summary. What do you notice? Tell your partner what it represents.

ELA CONNECTION

These suggested strategies address the Common Core State Standards for ELA for literacy.

RST 1: Cite evidence to support analysis of science texts.

RST 4: Determine the meaning of symbols, key terms, and other domain-specific words and phrases.

RST 6: Analyze the author's purpose in providing an explanation in a text.

Aerobic Cellular Respiration

No living thing can survive without energy. The trillions of cells in your body need energy to grow, reproduce, and respond to the environment.

All multicellular organisms, including plants, mushrooms, insects, and snails, must supply every cell with energy. And every single-celled organism, such as bacteria, must obtain energy. **Ⓐ**

Humans get energy from food. Food molecules, such as **glucose**, store potential energy in their chemical bonds. Cells can't use that potential energy directly. It is gradually released during a series of chemical reactions that require oxygen. Those reactions are **aerobic cellular respiration**. Here is the chemical summary. **Ⓑ**

$6O_2 + C_6H_{12}O_6 \rightarrow 6CO_2 + 6H_2O +$ *usable energy*
oxygen glucose   carbon   water **Ⓒ**
                    dioxide

Food tastes good, but that's not really why we eat. We get energy to live from the food we eat.

134   Full Option Science System

Part 2: Aerobic Cellular Respiration

Fats and proteins are also sources of energy-rich chemical bonds. Fats are used primarily for energy. Proteins provide raw materials for making muscle tissue, hormones, and other compounds and tissues. They can also be used for energy if sugar or fat is not available.

Aerobic cellular respiration occurs in every cell of almost every living thing. How do cells get the oxygen and glucose needed for aerobic cellular respiration?

Breathing and aerobic cellular respiration are related, but they are not the same.

Did You Know?
We often associate respiration with breathing. Aerobic cellular respiration, however, refers to the energy transfer processes in cells.

Insects, mushrooms, and every other living thing need energy to carry out life processes. The energy is released and used in the organism's cells.

Investigation 2: Supporting Cells 51

Cite evidence: Tell students to reread this page and discuss the question at the end. Then, have them continue reading the next two pages to confirm their understanding. [Humans breathe in oxygen from the air, which is carried from the lungs to the cells through the bloodstream. The digestive system breaks down food into sugars.]

Discuss the Did You Know? Have students discuss this statement with a partner: Respiration refers to breathing, bringing air into our lungs. Cellular respiration refers to the chemical processes that make usable energy in our cells.

ELA CONNECTION
This suggested strategy addresses the Common Core State Standards for ELA for literacy.

RST 1: Cite evidence to support analysis of science texts.

CROSSCUTTING CONCEPTS
Energy and matter

Human Systems Interactions Course—FOSS Next Generation   135

INVESTIGATION 2 – Supporting Cells

Understanding of words: Ask students what new concept is introduced in this section [Photosynthesis]. Have them share what they know about photosynthesis with a partner.

Research projects to answer a question: Ask students if they are familiar with the term "fermentation." Have a student read this section aloud. If students seem interested in finding out more, you might assign a student or two the task of researching this topic and presenting their findings to the class.

ELA CONNECTION

These suggested strategies address the Common Core State Standards for ELA for literacy.

WHST 7: Conduct short research projects to answer a question.

L 5: Demonstrate understanding of word relationships and nuances in word meaning.

Oxygen

Humans breathe in oxygen from the atmosphere. Blood picks up the oxygen in our lungs and carries it to the cells. There it becomes available for aerobic cellular respiration.

Plants don't breathe. So how do they get oxygen? Plants produce oxygen as a waste product during **photosynthesis**. Plant cells use some of that oxygen to perform aerobic cellular respiration.

Did You Know?

Some single-celled organisms live in places without oxygen. They use fermentation, which does not need oxygen, to produce energy. Our cells can use this process, too. It kicks in during intense exercise. But fermentation does not provide as much usable energy as aerobic cellular respiration.

Why is running an aerobic exercise? Because more physical activity requires more energy, and more energy requires more oxygen getting to the cells.

136   Full Option Science System

Part 2: Aerobic Cellular Respiration

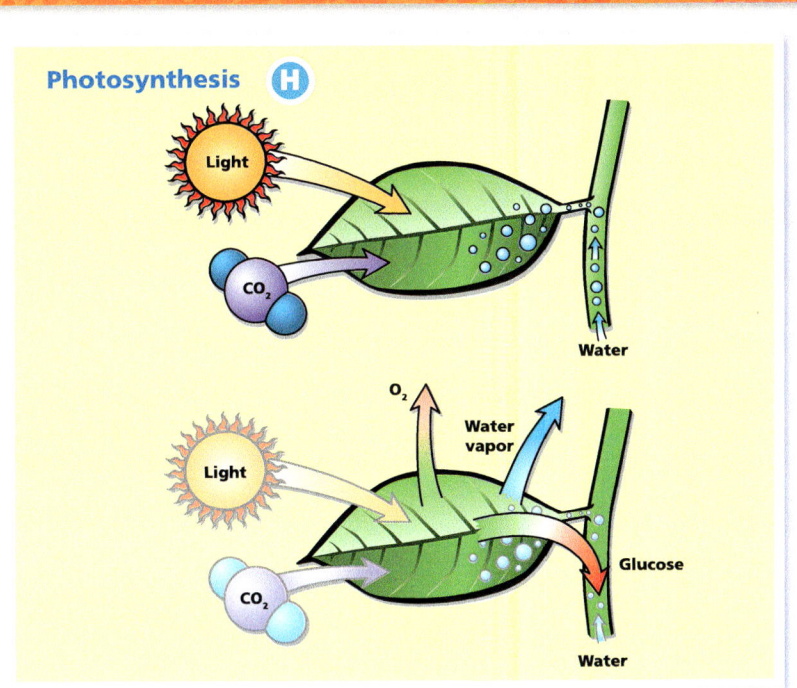

Photosynthesis

Plants use light energy from the Sun to turn carbon dioxide and water into sugar. Plants use the sugar for food and can use some of the oxygen for aerobic cellular respiration. The oxygen not used by the plant cells is released into the air.

Sugars

Some organisms can make their own glucose. They capture energy from the Sun, using a substance called chlorophyll. They use the Sun's energy to turn carbon dioxide gas and water into energy-rich sugars and oxygen. This process is photosynthesis. Here is a chemical summary of photosynthesis.

$$6CO_2 + 6H_2O + \text{light energy} \rightarrow C_6H_{12}O_6 + 6O_2$$

carbon dioxide · water · glucose · oxygen

The energy that originated as solar energy is now stored in the bonds of glucose molecules. It can be released as usable energy by aerobic cellular respiration. The cells of the photosynthetic organism can use this chemical energy to run their own activities.

Photosynthesis occurs mainly in the green leaves of plants.

Investigation 2: Supporting Cells · 53

Integrate technical information: Focus students' attention on the diagram. Have them take turns explaining the process of photosynthesis to each other using the model. Check for understanding using these questions.

➤ *What are the products of photosynthesis?* [Oxygen and glucose.]

➤ *Where does the energy come from to change carbon dioxide and water into oxygen and glucose?* [The Sun.]

Integrate technical information: Have students compare the chemical summary of photosynthesis with that of aerobic cellular respiration. Have them describe what is the same and what is different.

SCIENCE AND ENGINEERING PRACTICES

Constructing explanations

ELA CONNECTION

This suggested strategy addresses the Common Core State Standards for ELA for literacy.

RST 7: Integrate technical information expressed in words in a text with a version of that information expressed visually.

Human Systems Interactions Course—FOSS Next Generation

INVESTIGATION 2 – Supporting Cells

Determine the central idea: Tell students to look at the table and to compare the two processes. Ask students how photosynthesis and aerobic cellular respiration are related. [Photosynthesis uses the Sun's energy to make the substances that are used by cellular respiration to make usable energy for cells.]

Have students discuss the Think Question.

TEACHING NOTE

Energy originally comes from the Sun. But even in plant cells, that energy must be turned into usable energy for cells through aerobic cellular respiration. In other words, plant cells cannot use light energy directly for life processes.

ELA CONNECTION

This suggested strategy addresses the Common Core State Standards for ELA for literacy.

RST 2: Determine the central ideas or conclusions of a text; provide an accurate summary.

Humans can't make their own food. We eat plants or other animals that have eaten plants. Our digestive system breaks down food into sugar, protein, and fat molecules. Those energy-rich molecules move into the bloodstream from the small intestine, and travel to the cells. There, the food molecules become available for aerobic cellular respiration.

The cells break the bonds in the sugar molecule. Energy transfers to a usable form that cells (and consequently the body) can use to perform all the work they need to do.

Products

What are the products of aerobic cellular respiration? Carbon dioxide and water. These products leave the cells and are transported out of the body.

Summary

Aerobic cellular respiration is critical for life on Earth. The energy obtained from the breakdown of sugars, such as glucose, that are made by photosynthesis is what runs our cells and our bodies.

Think Question

A student said, "Plants do photosynthesis. Animals do aerobic cellular respiration." Do you agree or disagree with this statement? Explain your thinking.

Comparing Processes

	Aerobic cellular respiration	Photosynthesis
Occurrences	In cells of all organisms	Only in plants, algae, and some bacteria
Function	To break down food, making energy available to cells	To transfer energy from the Sun to chemical bonds
Reactants	Glucose and oxygen	Carbon dioxide, water, and light energy
Products	Carbon dioxide, water, and usable energy for cells	Glucose and oxygen

54

138 Full Option Science System

Part 2: Aerobic Cellular Respiration

16. Complete Frayer model for aerobic cellular respiration

When finished with the reading, have students complete the rest of their Frayer model. Have them fill in the "non-example" with a description of photosynthesis. Or, they can compare the similarities and differences between the processes.

Definition	Characteristics
The process almost all cells use to turn food molecules and oxygen into energy to live and grow	• requires oxygen • occurs in cells, both plant and animal cells • carbon dioxide and water are the waste products
Example	**Non-example**
In the human body, cells use glucose from the digestive system and oxygen from the respiratory system to obtain usable energy	Photosynthesis— plants capture energy from the Sun and use it to turn carbon dioxide and water into glucose and oxygen

(center: aerobic cellular respiration)

17. Review vocabulary
Take a few moments to review the vocabulary developed in this part. Have students update their vocabulary indexes if they haven't already done so.

18. Answer the focus question
Students should summarize what they have learned to respond to the focus question.

▶ *How does the energy in food become energy that cells can use?*

ELA CONNECTION

This suggested strategy addresses the Common Core State Standards for ELA for literacy.

L 6: Acquire and use academic and domain-specific words.

 aerobic cellular respiration
glucose

TEACHING NOTE

Students may answer simply that cells use aerobic cellular respiration to turn food and oxygen into usable energy. If so, ask for greater detail, including how the resources get to the cells.

Human Systems Interactions Course—FOSS Next Generation 139

INVESTIGATION 2 – Supporting Cells

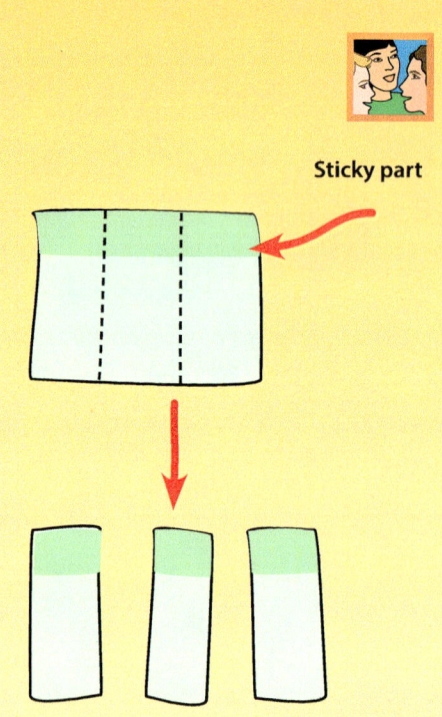

Sticky part

DISCIPLINARY CORE IDEAS

PS3.D: Energy in chemical processes and everyday life

LS1.A: Structure and function

LS1.C: Organization for matter and energy flow in organisms

WRAP-UP

19. Review notebook entries

Distribute one self-stick note to each student. Ask students to cut the note into three pieces, making sure that each piece has a sticky end.

Ask students to take 3 minutes to look back through their notebook entries to find the three most important things they learned in Investigation 2. They should tag those pages with the three self-stick notes. They might use a highlighter or colored pencils to call out the key points.

After students tag their three big ideas (no more), give them 3 minutes for a group discussion to share the ideas they value most in this investigation. They should also look back at the big ideas for Investigation 1.

Here are the big ideas that should come forward for both investigations.

- The human body is made of systems, which are made of organs, which are made of tissues, which are made of cells.
- All the human organ systems interact in order for a human to live and carry out life functions. The most important function is servicing cells.
- In a human, the circulatory system pumps blood, which carries resources to each cell and carries away waste.
- Cells use glucose and oxygen, provided by the digestive and respiratory systems, to provide usable energy for the cell. This process is aerobic cellular respiration.
- Carbon dioxide and water are the waste products of aerobic cellular respiration and are removed by the respiratory and excretory systems.
- Modeling systems is useful to describe processes and construct explanations.

Another review strategy is to have students work together to make a concept map. Make sets of the vocabulary words from investigations 1 and 2 on self-stick notes for each group of four students (or have students make them on their own). Each group takes the set of words and decides how to organize them on a big piece of paper. Once the group decides how the words are related, they draw lines connecting the words. On the lines they write words or phrases that describe or explain how the words/concepts are connected. When students have finished their concepts maps, post them on the wall so groups can compare their thinking.

Part 2: Aerobic Cellular Respiration

SESSION 3 45–50 minutes

20. Assess progress: I-Check

Administer *Investigations 1–2 I-Check*, asking students to respond to the items on paper with pencil or online on FOSSmap. Students should independently answer the questions. When taking the I-Check, students should not use their notebooks, but the notebooks are a good tool to use when students later reflect on their answers.

SESSION 4 45–50 minutes

21. Discuss I-Check results

Code the I-Checks items, but do not make any marks on student responses. Answer sheets and coding guides can be found in the Assessment chapter. You can record student results on the *Assessment Record* or download spreadsheets from FOSSmap for recording. Note important points about the items to review with students.

Return the I-Checks to students. Use self-assessment strategies as described in the Assessment chapter for each item to facilitate reflection and clarify student thinking.

Human Systems Interactions Course—FOSS Next Generation

INVESTIGATION 2 – Supporting Cells

EXTENDING *the Investigation*

- **Research aerobic cellular respiration**
  What is fermentation and how is it related to aerobic cellular respiration?

- **Research bacteria**
  Some bacteria use other substances, like sulfur, instead of oxygen during cellular respiration. Research one of those organisms to see what else can be used and where those organisms live.

- **Research mitochondrial disease**
  Mitochondrial disease is a problem with cells producing energy. How does mitochondrial disease occur, and which body systems does it affect most often?

INVESTIGATION 3 – The Nervous System

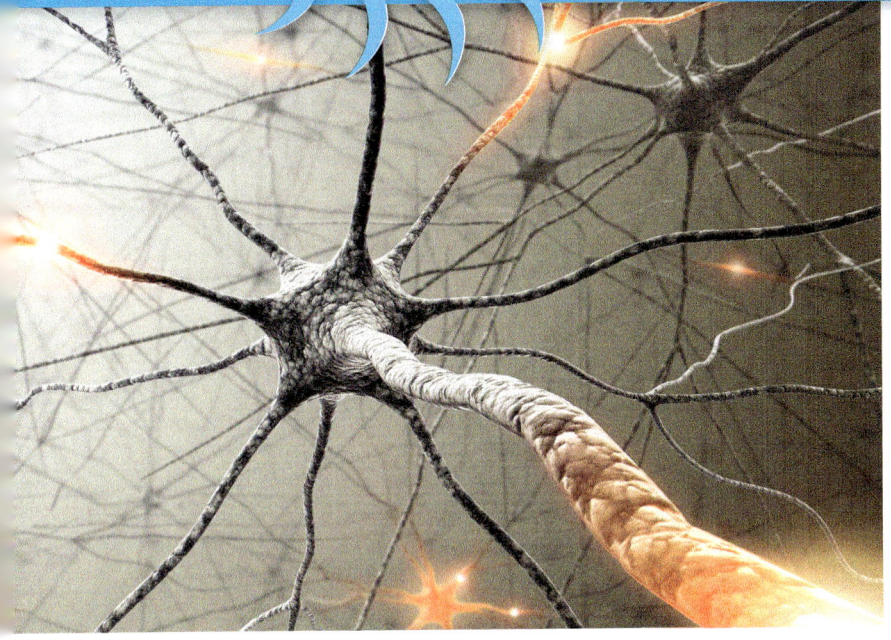

Part 1
The Sense of Touch **156**

Part 2
Sending a Message **175**

Part 3
Other Senses **199**

Part 4
Learning and Memory **216**

PURPOSE

In *The Nervous System*, students explore the different senses to understand how humans gather information from the environment. They engage in a "neuron relay" to model how sensory information travels to the brain for processing and how information returns to the body for action. Finally, students turn their attention to their own learning and memory formation.

Content

- Sensory receptors respond to an array of mechanical, chemical, and electromagnetic stimuli.
- Sensory information is transmitted electrically to the brain along neural pathways for processing and response.
- Neural pathways change and grow as information is acquired and stored as memories.

Practices

- Develop a model for the action of a neural pathway.
- Gather and interpret data on sensory stimuli and responses.
- Experience and describe learning and memory formation.

Science and Engineering Practices
- Developing and using models
- Planning and carrying out investigations
- Analyzing and interpreting data
- Constructing explanations
- Engaging in argument from evidence
- Obtaining, evaluating, and communicating information

Disciplinary Core Ideas
LS1: How do organisms live, grow, respond to their environment, and reproduce?
LS1.A: Structure and function
LS1.D: Information processing

Crosscutting Concepts
- Patterns
- Cause and effect
- Scale, proportion, and quantity
- Systems and system models
- Structure and function

INVESTIGATION 3 – The Nervous System

	Investigation Summary	Time	Focus Questions and Practices
PART 1	**The Sense of Touch** Students think about how humans sense the environment around them and then turn their attention to the sense of touch. They compare touch sensitivity between fingertips and knuckles to learn about pressure receptors and receptive fields.	Active Inv. 2 Sessions	**How does the sense of touch work in humans?** **Practices** Developing and using models Planning and carrying out investigations Analyzing and interpreting data Constructing explanations Engaging in argument from evidence Obtaining, evaluating, and communicating information
PART 2	**Sending a Message** Students consider the stimulus/response phenomenon. They develop a model to explain how messages are transmitted along neurons and across synapses, to and from the brain.	Active Inv. 3 Sessions	**How do messages travel to and from the brain?** **Practices** Developing and using models Analyzing and interpreting data Constructing explanations Engaging in argument from evidence Obtaining, evaluating, and communicating information
PART 3	**Other Senses** Students explore the sense of smell by identifying scents, and the sense of sight by testing reaction time. They read about chemical receptors and photoreceptors and consider how their eyes are designed to interpret electromagnetic information.	Active Inv. 3 Sessions	**How are the senses alike and how are they different?** **Practices** Analyzing and interpreting data Constructing explanations Obtaining, evaluating, and communicating information
PART 4	**Learning and Memory** Students use mirror drawing to explore the connection between hand–eye coordination, learning, and memory. They use various combinations of sensory input to memorize lists of objects. They look for patterns to determine strategies for improving short-term memory.	Active Inv. 3 Sessions Assessment 3–4 Sessions	**How do humans learn and form memories?** **Practices** Analyzing and interpreting data Constructing explanations Obtaining, evaluating, and communicating information

* A class session is 45–50 minutes.

At a Glance

Content Related to DCIs	Literacy/Technology	Assessment
• Sensory receptors respond to an array of mechanical, chemical, and electromagnetic stimuli.	**Science Notebook Entry** *Touch* **Science Resources Book** "Sensory Receptors" "Touch" "Hearing" **Online Activities** "Touch Menu: Touch Receptors" "Touch Menu: 3D Finger"	**Embedded Assessment** Science notebook entry
• Sensory information is transmitted electrically to the brain along neural pathways for processing and response.	**Science Notebook Entry** *Neural Pathways* **Science Resources Book** "Brain Messages" "Neurotransmission" **Online Activities** "Brain: Synapse Function" "Brain: Neuron Growth"	**Embedded Assessment** Response sheet
• Sensory receptors respond to an array of mechanical, chemical, and electromagnetic stimuli. • Sensory information is transmitted electrically to the brain along neural pathways for processing and response.	**Science Notebook Entry** *Smell* *Reaction-Timer Results* **Science Resources Book** "Sensory Receptors" "Smell and Taste" "Sight" **Online Activities** "Smell Menu" "Vision Menu" "Reaction Timer"	**Embedded Assessment** Science notebook entry
• Neural pathways change and grow as information is acquired and stored as memories.	**Science Notebook Entry** *Mirror Drawing* *"How Memory Works" Video Questions* **Science Resources Book** "Memory and Your Brain" **Video** *How Memory Works*	**Embedded Assessment** Science notebook entry **Benchmark Assessment** Investigation 3 I-Check Posttest **NGSS Performance Expectations addressed** MS-LS1-3; MS-LS1-8

Human Systems Interactions Course—FOSS Next Generation

INVESTIGATION 3 – The Nervous System

SCIENTIFIC and Historical Background

Just about every human thought, emotion, and physical action is initiated and controlled by the brain. The brain is a 1.5 kilograms (kg) mass of some 60 billion neurons that are in communication with each other at all times. It is marvelously complex, responding to the external and internal environment, sending and receiving messages via neural pathways, proposing actions and comparing them to previous actions, building memories and constructing knowledge. It has the awesome ability to redesign itself as it goes about its business, adding new connections between parts. And each brain is different from every other in its detail and finish. Hence the dramatic diversity in humanity.

The brain interacts with the rest of the body; its communication network can be compared to a power grid that is never at rest. Even while we sleep, the nervous system constantly receives and interprets signals from the surrounding world and the internal organs to maintain homeostasis, monitoring and regulating the body's functions. An engaging entrance into the study of the nervous system is via the sense of **touch**.

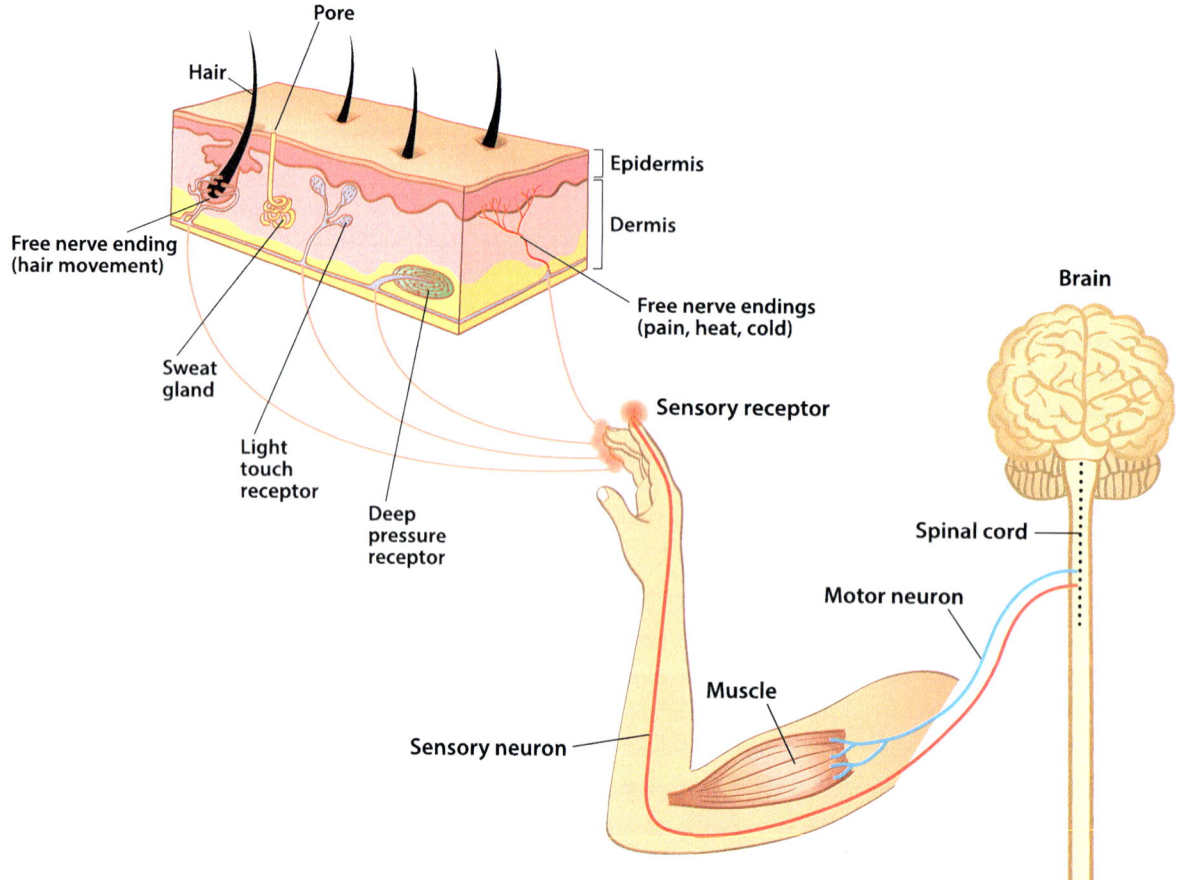

Scientific and Historical Background

How Does the Sense of Touch Work in Humans?

We acquire information about our environment through our senses. Senses pick up data about the world around us, and pass them on to the brain. The brain compares the data to information that is already cataloged, and initiates an appropriate action or **response**.

Touch is an essential part of the human sensory experience. The sense of touch is a catchall phrase that includes three major types of **stimuli**.

- Pressure is a mechanical stimulation of the skin. The skin has several kinds of **mechanoreceptors** that respond to different kinds of pressure, including fluttering light touch (tickle, caress), deep steady indentations of the skin, and vibrations.
- Pain (nociception) affects different receptors for sharp stabbing pain, deep chronic aches, and persistent acute pain caused by damage resulting from burning or freezing of the skin.
- Temperature affects different receptors for hot and cold.

Each sensory receptor has a **receptive field**, or area from which it gathers information. The number, spacing, and size of the receptors are what determine the ability to discriminate between two touches. A doctor can use the information from a two-point discrimination test to determine how finely innervated an area of skin is. The results can be compared to normative data to find out if a patient's sensory nerve function is normal.

Receptors for pressure are most densely packed in the fingertips and mouth, and least densely spaced in areas of the back and legs. Our ability to discriminate between two points varies at least 20-fold in different areas of our bodies. This variation can be portrayed in a representation of a person with the parts of the body scaled to correspond to the density of their sensory receptors, the homunculus.

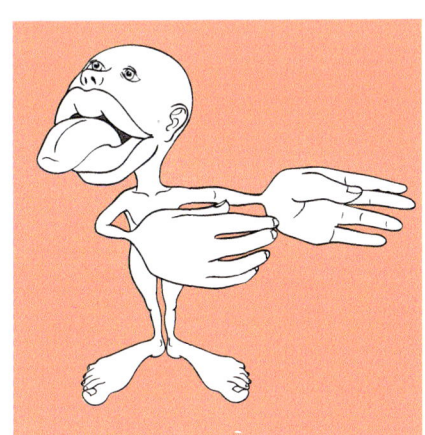

The **sense receptors** for hearing are also mechanoreceptors. These mechanoreceptors are hair cells in the fluid-filled cochlea, which respond to vibrations that are transmitted by the ear drum and bones in the ear. The hair cells send signals to the brain for processing.

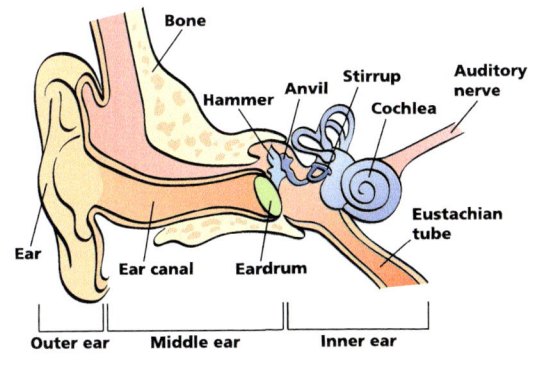

The human ear

Human Systems Interactions Course—FOSS Next Generation

INVESTIGATION 3 – The Nervous System

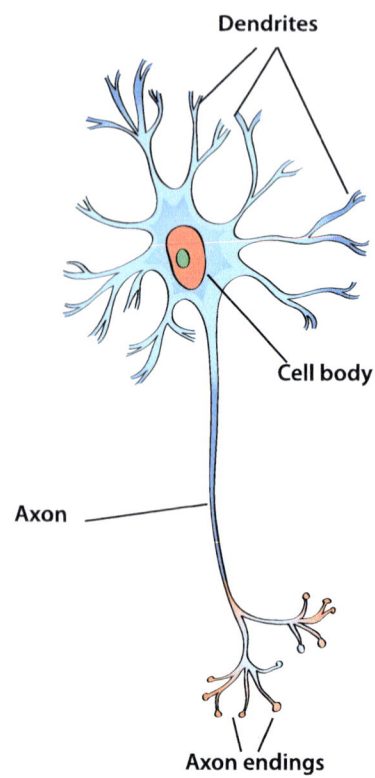

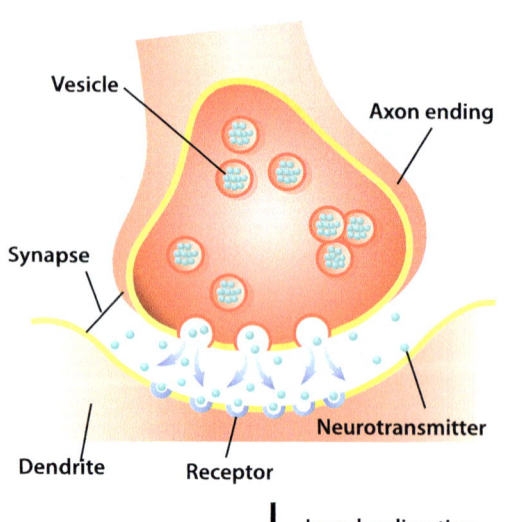

How Do Messages Travel to and from the Brain?

Sensory input from mechanoreceptors and other sources can stimulate responses or memory formation. Data processing occurs in the brain, not at the site of the sensory input. The body needs an efficient system to transmit data from sensory receptors to the brain and back to the appropriate muscles to take action. All this data transmission has to happen quickly enough so that a person can, more often than not, withdraw a hand from a hot object, jump out of oncoming traffic, or avoid any number of other potentially harmful situations before damage to the body occurs. Human existence relies on this rapid data transmission and processing system.

The "wires" of the system are **nerves,** which are made of cells called **neurons**. Neurons are specially designed to transmit information from place to place throughout the body.

The cell body is the operational center of the neuron, and contains the nucleus. Extending from the cell body are numerous branches called dendrites, which receive incoming messages. The messages are transmitted to the dendrites, and travel through the cell body and into a long thin extension of the cell called the axon. At the end of the axon, the message passes to the dendrites(s) of another neuron, and the message continues on its way.

The message itself is an electric impulse. The electric impulse travels through the neuron to the axon endings, where it is transmitted chemically to the dendrites of the next neuron.

Synapses. Between the axon endings of one neuron and the dendrites of the next is a small space called a **synapse**. The end of the axon contains tiny vesicles filled with chemicals called **neurotransmitters**. When an electric impulse arrives at the axon terminus, neurotransmitters are released. They travel across the synapse and bind to receptors on the cell membrane of the next neuron's dendrites. When the neurotransmitters trigger the dendrite receptors, they open gates that allow charged ions to surge through the next neuron. The charged ions continue the electric message in the receiving dendrite, and the impulse races on to the next neuron. This combination of electric impulses and chemical neurotransmitters is why communication in the nervous system is termed electrochemical.

Scientific and Historical Background

Neurons. There are three different kinds of neurons: sensory, motor, and interneurons. Sensory neurons carry messages from the sensory receptors toward the brain. Some sensory neurons have specialized receptor bodies as nerve endings in place of dendrites. Receptors in these specialized sensory neurons do not respond to neurotransmitters; they respond to the particular stimulus they are designed to receive (light, pressure, sound waves, and so on). These sensory neurons carry messages to the spinal cord, where the message is transmitted to interneurons, which act as intermediaries between sensing and acting.

Interneurons carry the message up to the brain, where it is processed by a host of other interneurons. The interneurons either generate a signal for action and/or form memories. The outgoing message is carried to muscles or glands. Motor neurons that connect to muscle cells end in synapses with the muscle cells themselves, telling them to contract.

How Are the Senses Alike and How Are They Different?

Now that we have an inkling about how the nervous system works, we can explore the mechanisms of other senses: **vision**, **smell**, and taste.

Vision. Visible light waves along the electromagnetic spectrum enter the eye through the pupil and stimulate **photoreceptors** located in the retina in the back of the eye. The photoreceptors, rod and cone cells, respond differently to light. Rods respond to dim light and are concentrated away from the center of the retina. Cones respond to bright light, providing information about color and detail. Cones are concentrated in the center of the retina. Each photoreceptor cell converts the light stimulus into electric impulses, which travel through the optic nerve to the brain for processing.

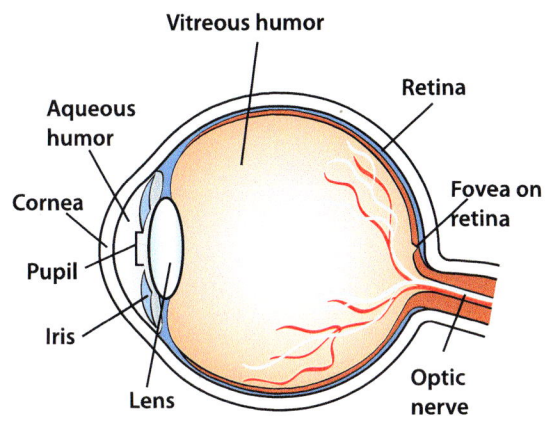

The time it takes for a person to see something, the message to travel to the brain, the brain to figure out how to respond, and the message to travel to a muscle for immediate action is called **reaction time**. Reaction time is surprisingly complex. It can vary greatly across different tasks as well as within the same task under different conditions.

Smell and taste. Smell and taste receptors are **chemoreceptors**. Chemoreceptors respond to chemicals. For the sense of smell, the stimulus is a wide array of chemicals that evaporate in the air and then dissolve in the mucus of the olfactory surface at the top of the nasal cavity. Chemoreceptors are located on that surface and respond to one or a combination of the basic odor types. The chemical molecules fit into specific shapes on the receptors and stimulate electric messages that head off to the olfactory bulb in the brain.

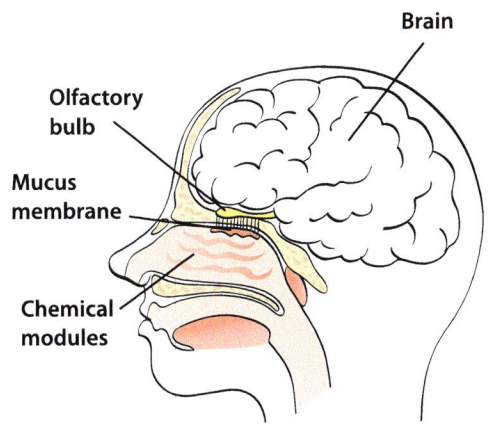

Human Systems Interactions Course—FOSS Next Generation 149

INVESTIGATION 3 – The Nervous System

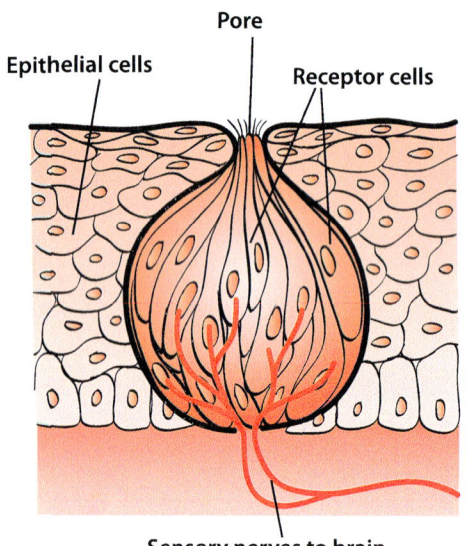

A taste bud

Taste chemoreceptors, on the other hand, respond to chemicals that are dissolved in the saliva and come into contact with taste buds located on the tongue. The receptors respond to one type of "taste chemical," (salt, sweet, bitter, sour, umami or savory). When stimulated, the taste receptors send electric signals to the brain for processing. The smell of food contributes to the final sensation of taste.

How Do Humans Learn and Form Memories?

The brain is constantly changing, influenced by its interactions with the world. Sensory experiences change neural pathways as the connections between neurons are strengthened, weakened, or newly created. And stem cells in the brain (neural stem cells) can generate new neurons, even into adulthood. Many of these new cells originate in regions of the brain associated with **learning** and **memory**.

Human neurons form during fetal development at the astonishing range of 250,000 neurons per minute over the 9 months of pregnancy. In the first few years of life, neuron formation slows, but the connections between neurons, known as neural networks, grow as infants learn and experience the world. These intricate patterns of connections somehow encode our personalities.

Learning is defined as the acquisition of new knowledge and skills. The process of thinking about our thinking is called **metacognition** and can help us understand how we learn.

Memory refers to the storage and recall of information, including past experiences, knowledge, skills, and thoughts. The various types of memory include sensory memory, short-term memory, working memory, and long-term memory. These memory systems interact continuously to give rise to the phenomenon of memory as we experience it in our everyday lives. Most types of memory appear to be stored in the **cerebral cortex**, the outer layer of the cerebrum of the brain.

There has been much research into how memories form in our brains. In the 1950s, a cellular mechanism for the formation of memories was proposed. This hypothesis was that learning and memory rely on the strengthening of connections between neurons. At the same time, another scientist suggested that simultaneous activity among neurons that are already actively connected in a network can selectively strengthen their synaptic connections. And conversely, if connections are not active, the network may weaken or stagnate. These ideas have been confirmed with specific laboratory studies.

Scientific and Historical Background

According to the National Institutes of Health, it appears that short-term memory involves a temporary strengthening of connections between already existing synapses, allowing the network to be sensitized to subsequent messages. And if an experience is significant enough or if it is repeated enough times, it stimulates a biochemical process that causes specific genes to stimulate the production of synapse-strengthening proteins. These proteins permanently strengthen an existing network.

Those networks inform who we are as unique individuals. All of this happens in the brain, that marvelous and complex instrument located inside our skull.

INVESTIGATION 3 – The Nervous System

NGSS Foundation Box for DCI

LS1.A: Structure and function
- In multicellular organisms, the body is a system of multiple interacting subsystems. These subsystems are groups of cells that work together to form tissues and organs that are specialized for particular body functions.

LS1.D: Information processing
- Each sense receptor responds to different inputs (electromagnetic, mechanical, chemical), transmitting them as signals that travel along nerve cells to the brain. The signals are then processed in the brain, resulting in immediate behaviors or memories.

Vocabulary
Cerebral cortex
Chemoreceptor
Learning
Mechanoreceptor
Memory
Metacognition
Nerve
Neuron
Neurotransmitter
Photoreceptor
Reaction time
Receptive field
Response
Sensory receptor
Smell
Stimulus
Synapse
Touch
Vision

TEACHING AND LEARNING *about the Nervous System*

Developing Disciplinary Core Ideas (DCI)

Your students are at the right age to get to know the brain on an intimate basis. Their brain is their operating system, central information processing center, and the very essence of who they are. The fantastically complex electrochemical activities are supported and nourished by the chemical environments and intellectual stimulation they provide for it. Middle school students are mature enough to get their arms around these concepts. Are they mature enough to act in their own best interest? That's a tough one.

It is our belief that sound scientific information coupled with experiences that guide students to critical thinking and problem solving will prepare students for the life-molding decisions that face them at this time in their lives. If they have an understanding about the structures and functions of their own nervous system and the mechanisms that enable them to sense and respond to their environment, they might be better able to care for this system and use it wisely.

Of all human endeavors, the desire to understand ourselves is paramount. Learning about their own ability to process sensory information and form memories helps students reflect on themselves as lifelong learners and develop a sense of the brain and its remarkable capacity as a flexible, growing network. Students will start to move past any self-perceived limitations as they redefine themselves as organisms capable of self-reflection, growth, and change. Research shows that students who develop a growth mindset are more likely to see challenges as opportunities, a perspective that is invaluable not just in the classroom but for the many challenging situations students will face outside of the classroom, in their school years and beyond.

The experiences students have in this investigation contribute to the disciplinary core ideas **LS1.A: Structure and function** and **LS1.D: Information processing**.

Teaching and Learning about the Nervous System

Engaging in Science and Engineering Practices (SEP)

In this investigation, students engage in these practices.

- **Developing and using models** about receptive fields for pressure sensory receptors in hands to predict phenomena and to generate data to test ideas about those models; and about neural-message transmission from a muscle to the brain and back to the muscle to describe phenomena of reaction time.

- **Planning and carrying out investigations** dealing with pressure sensory receptors on different parts of the hand and reaction time to collect data to serve as the basis for evidence to answer scientific questions about information processing in humans.

- **Analyzing and interpreting data** on pressure sensory receptors on different parts of the hand to provide evidence to support a model describing the information processing; on symptoms related to sensory problems to construct an explanation for the medical condition; and on a person's reaction time in bright and dim light and integrate that with information on functions of different kinds of photoreceptors in the eye.

- **Constructing explanations** by applying scientific ideas about pressure sensory receptors in the hands, about stimulus–response actions mediated by the human nervous system, and about the response thresholds of photoreceptors in the eye.

- **Engaging in argument from evidence** about models that describe pressure sensor receptor fields on different parts of the hands, and about what can go wrong in the nervous system to result in a particular condition or phenomenon. Respectfully provide and receive comments about explanations by citing data or text to support the case.

- **Obtaining, evaluating, and communicating information** from written text and other visual media integrated with personal experiences to construct explanations about how humans organs and systems gather, process, and respond to information from the environment.

NGSS Foundation Box for SEP

- **Develop and/or use a model** to predict and/or describe phenomena.
- **Develop and/or use a model to generate data** to test ideas about phenomena in natural or designed systems, including those representing inputs and outputs, and those at unobservable scales.
- **Collect data to serve as the basis for evidence** to answer scientific questions or test design solutions under a range of conditions.
- **Analyze and interpret data** to provide evidence for phenomena.
- **Apply scientific ideas, principles, and/or evidence** to construct, revise, and/or use an explanation for real-world phenomena, examples, or events.
- **Respectfully provide and receive critiques** about one's explanations, procedures, models, and questions by citing relevant evidence and posing and responding to questions that elicit pertinent elaboration and detail.
- **Construct, use, and/or present an oral and written argument** supported by empirical evidence and scientific reasoning to support or refute an explanation or a model for a phenomenon (or a solution to a problem).
- **Integrate qualitative and/or quantitative scientific and/or technical information** in written text with that contained in media and visual displays to clarify claims and findings.
- **Communicate scientific and/or technical information** (e.g., about a proposed object, tool, process, system) in writing and/or through oral presentations.

INVESTIGATION 3 – The Nervous System

NGSS Foundation Box for CC

- **Patterns:** Patterns in rates of change and other numerical relationships can provide information about natural and human-designed systems. Patterns can be used to identify cause-and-effect relationships.
- **Cause and effect:** Cause-and-effect relationships may be used to predict phenomena in natural or designed systems.
- **Scale, proportion, and quantity:** The observed function of natural and designed systems may change with scale. Phenomena that can be observed at one scale may not be observable at another scale.
- **Systems and system models:** Systems may interact with other systems; they may have subsystems and be a part of larger complex systems. Models are limited in that they only represent certain aspects of the system under study.
- **Structure and function:** Complex and microscopic structures and systems can be visualized, modeled, and used to describe how their function depends on the shapes, composition, and relationships among its parts; therefore, complex natural and designed structures/systems can be analyzed to determine how they function.

Exposing Crosscutting Concepts (CC)

In this investigation, the focus is on these crosscutting concepts.

- **Patterns.** Patterns resulting from different strategies of learning a new skill can be used to improve learning.
- **Cause and effect.** Sensory input and output (stimulus and response) are cause-and-effect relationships that help us understand the functioning of the system.
- **Scale, proportion, and quantity.** Sensory systems can be observed and represented at many scales—system, organ, cell, cell structure. Each representation provides new understandings about the sensory system.
- **Systems and system models.** Sensory system models are useful tools to describe and study information processing in organisms. Sensory system models are limited in that they usually represent only certain aspects of the system.
- **Structure and function.** The structures and functions of the human nervous system can be modeled at many scales and used to describe how information processing and response occur.

Connections to the Nature of Science

- **Scientific knowledge assumes an order and consistency in natural systems.** Science assumes that objects and events in natural systems occur in consistent patterns that are understandable through measurement and observation. Science carefully considers and evaluates anomalies in data and evidence.
- **Science is a human endeavor.** Scientists and engineers rely on human qualities such as persistence, precision, reasoning, logic, imagination, and creativity. They are guided by habits of mind, such as intellectual honesty, tolerance of ambiguity, skepticism, and openness to new ideas. Advances in technology influence the progress of science, and science has influenced advances in technology.

Connections to Engineering, Technology, and Applications of Science

- **Interdependence of science, engineering, and technology.** Engineering advances have led to important discoveries in virtually every field of science, and scientific discoveries have led to the development of entire industries and engineered systems. Science and technology drive each other forward.

Teaching and Learning about the Nervous System

Conceptual Flow

The conceptual flow starts in Part 1 as students focus on the **nervous system** and think about how humans sense the environment. Students focus on the **sense of touch** and learn about the **mechanoreceptors in the skin that detect pressure, pain, and temperature**. They discuss models for **pressure receptive fields**, collect first-hand data, and engage in argument from evidence about alternative models. They read about hearing and the **mechanoreceptors involved in detecting vibrations**.

In Part 2, students analyze an example of stimulus and response and take part in a **simulation of a neural-message relay that involves sensory neurons, spinal-cord interneurons, brain neurons, and motor neurons**. They learn that sensory information is transmitted electrically to the brain along **neural pathways** for processing and response, and the transmission requires chemical **neurotransmitters to traverse the synapse between the neurons**.

In Part 3, students learning about **chemoreceptors involved in smell and taste**, and **photoreceptors (electromagnetic) involved in vision**. They explore sense of smell by matching mystery scents and gather information about how smell travels to the brain. They conduct a **reaction time activity** using vision and compare their reaction time in light and dark environments. They analyze the results and discuss strategies for improvement using their knowledge of the different kinds of photoreceptors in the retina of the eye.

In Part 4, students have a novel **learning** experience (mirror writing) and engage in metacognition about strategies to learn a new skill. They are presented with **memory tasks to gather information about how they best use their senses to remember**. Learning and memory are the results of **changes and growth in neural pathways**.

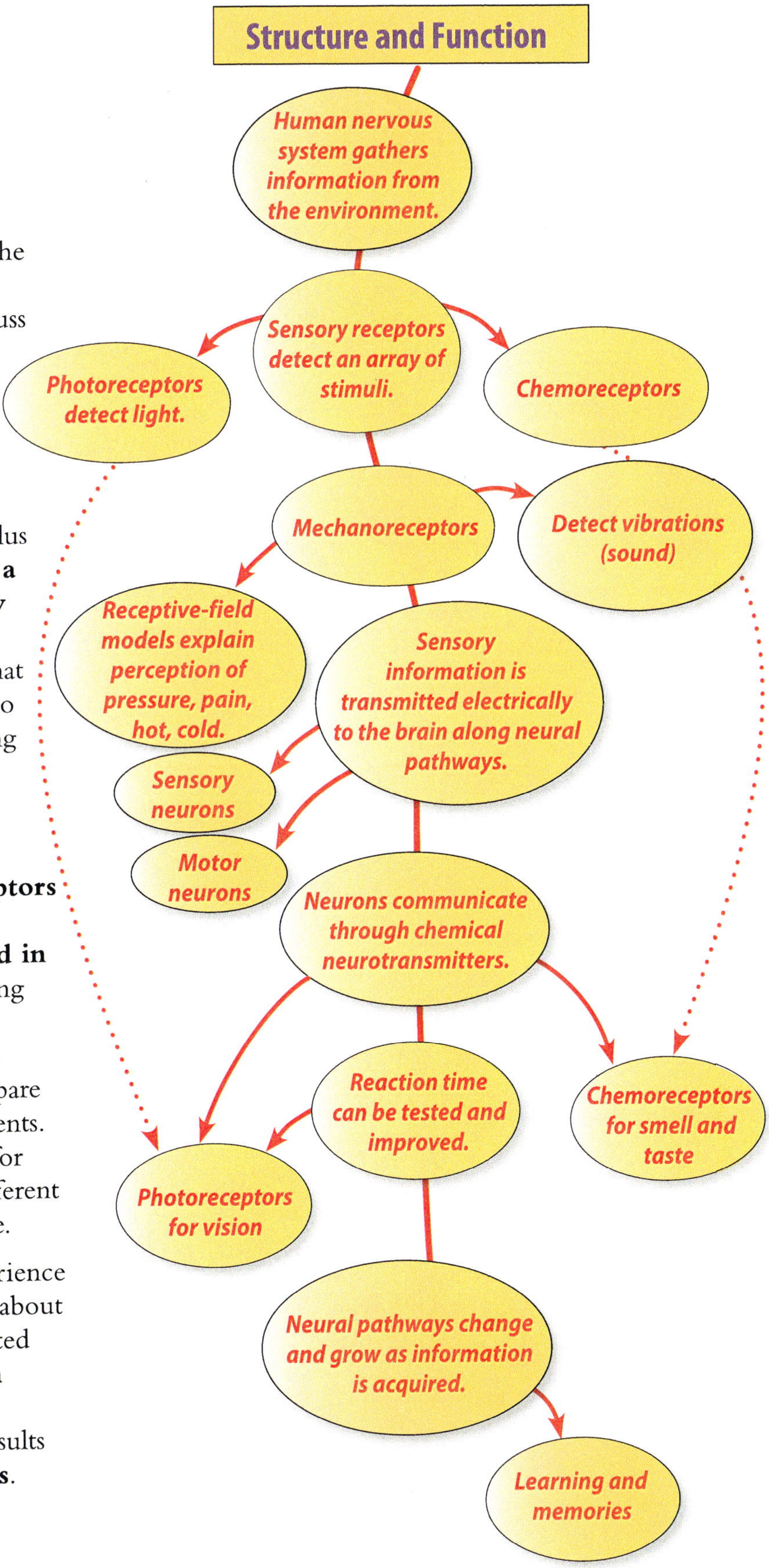

Human Systems Interactions Course—FOSS Next Generation

155

INVESTIGATION 3 – The Nervous System

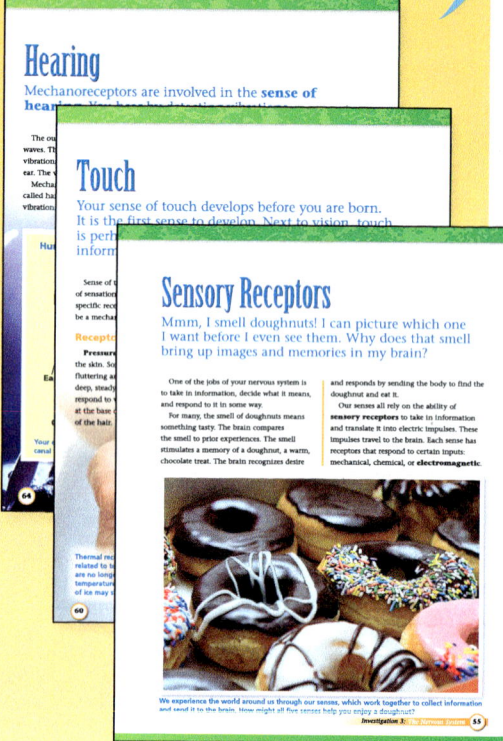

FOSS Science Resources

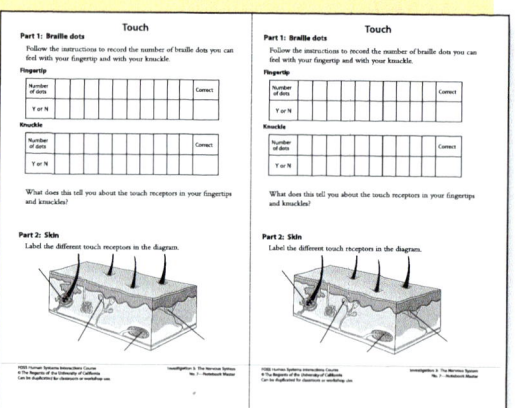

No. 7—Notebook Master

Teacher Master E

MATERIALS for
Part 1: The Sense of Touch

Provided equipment
For each student
- 1 *FOSS Science Resources: Human Systems Interactions*
- "Sensory Receptors"
- "Touch"
- "Hearing"

For each group
- 2 Braille strips
- 2 Paper clips, jumbo

Teacher-supplied items
For each group
- 2 Rulers
- 1 Computer with Internet access (optional)

For the class
- Rubbing alcohol (optional)
- Document camera (optional)

FOSSweb resources
For each student
- 1 Notebook sheet 7, *Touch*

For the class
- Online activity: "Touch Menu: Touch Receptors"
- Online activity: "Touch Menu: 3D Finger"

For the teacher
- Teacher master C, *Pressure Receptive-Fields Models A*
- Teacher master D, *Pressure Receptive-Fields Models B*
- Teacher master E, *Homunculus*
- Embedded Assessment Notes
- Teaching slides, 3.1

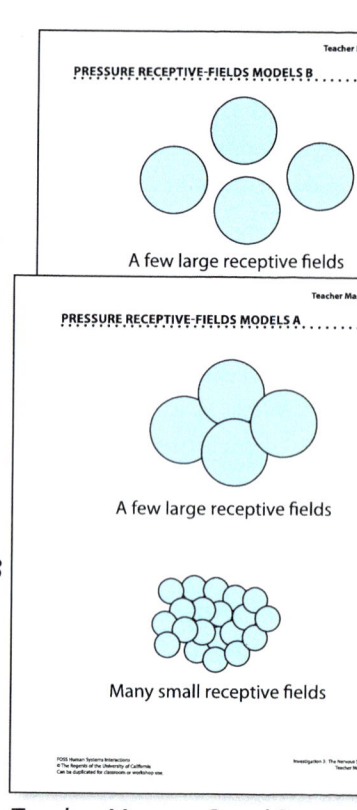

Teacher Masters C and D

Full Option Science System

Part 1: The Sense of Touch

GETTING READY for
Part 1: The Sense of Touch

Quick Start

Schedule	• Computers with Internet access, one for each group (optional) 2 active investigation sessions, including 2 readings
Preview	• Preview the FOSSweb Resources by Investigation for this part (such as printable masters, teaching slides, and multimedia) • Preview online activities: "Touch Menu: Touch Receptors," Step 6 "Touch Menu: 3D Finger," Step 6; Step 10 • Preview the reading: "Sensory Receptors," Step 7 [A] • Preview the reading: "Touch," Step 9 • Preview/plan reading for homework: "Hearing," Step 20
Print or Copy	**For each student** • Notebook sheet 7 **For the teacher** • *Embedded Assessment Notes*
Prepare Material	• Prepare braille strips [B]
Plan for Assessment	• Review Step 16, "What to Look For" in the notebook entry

▶ **NOTE**
Schedule computers for each group for Parts 1 and 3.

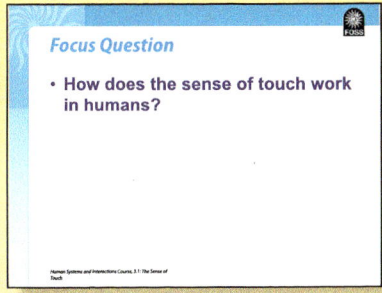

▶ **NOTE**
Preview the teaching slides on FOSSweb for this part.

Human Systems Interactions Course—FOSS Next Generation 157

INVESTIGATION 3 – The Nervous System

Preparation Details

A▶ Preview reading about sensory receptors

In Step 7 of Guiding the Investigation, students will read the first two pages of "Sensory Receptors" in *FOSS Science Resources*, the introduction and a page about mechanoreceptors. Students will return to this article in Part 3 of the investigation to read the sections on chemoreceptors and photoreceptors. For this part, the focus is on touch and hearing, both examples of mechanoreceptors.

B▶ Prepare braille strips

Students use plastic strips that have braille dots imprinted on them. Check the condition of the strips.

As these strips are to be used from year to year, you might want to clean them with rubbing alcohol after each class.

SENSE-MAKING DISCUSSION PLANNING GUIDE

Course	Human Systems Interactions	
Investigation: Inv. 3 The Nervous System		Part 1: The Sense of Touch
Guiding question: How do humans detect, process, and use information about the environment?		
Focus question: How does the sense of touch work in humans?		

NEXT GENERATION SCIENCE STANDARDS

Focus DCI(s)	Focus SEP(s)	Focus CCs
LS1.A: Structure and function LS1.D: Information processing	**Developing and using models**, Planning and carrying out investigations, **Analyzing and interpreting data**, Constructing explanations, **Engaging in argument**, Obtaining, evaluating, and communicating information	Systems and system models, Structure and function, Scale, proportion, and quantity

Questions and Intended Responses

Step 15

What to ask	➤ ***What data could you collect to develop evidence to support the overlapping or nonoverlapping model (for receptive fields)? (Step 15)*** ➤ What was your test? ➤ What was the data you collected? ➤ What model do you claim represents how pressure receptive fields are arranged in human skin? Summarize your evidence regarding both size and overlapping or nonoverlapping receptive fields.
What to listen for	• We used the paper clip testing technique to search for areas of skin where we can't feel pressure. • If we always felt the paper clip, that supports the overlapping model. • If there were places where we didn't feel the paper clip, that supports the nonoverlapping model because there are places that are not covered by receptive fields. • We thought about size in the prior test; if smaller receptive fields, we always feel the correct number; if larger, we can't always tell.
Scaffolding questions	➤ How are the overlapping and nonoverlapping models for pressure receptive fields different? ➤ If there were tests where you didn't feel the paper clip, what does that tell you? ➤ If you always felt the paper clip, what does that tell you?
Application questions	➤ Do you think the number of pressure receptors and model of pressure receptive fields are the same in all parts of the body that have skin? (Back, arms, legs, feet, etc.) (this is looking forward to Step 17) ➤ We thought about pressure receptive fields. What about other receptors in your fingers and knuckles? Might they follow the same model? ➤ Do you think pain receptor receptive fields might be large or small? What might be the pros and cons of large or small? ➤ If a receptive field is large, might it be easier or harder to accurately pinpoint where the stimulus is? (harder, anywhere within the field)

Part 1: The Sense of Touch

GUIDING *the Investigation*
Part 1: *The Sense of Touch*

SESSION 1

Students will . . .
- Focus on the sense of touch to find out how it works (Steps 1–3)
- Gather and discuss sensory receptor data on hand (Steps 4–6)
- Read about mechanoreceptors in "Sensory Receptors" (Steps 7, 8)
- Read and discuss "Touch" (Steps 9, 10)

SESSION 2

Students will . . .
- Discuss models for pressure receptive fields (Steps 11, 12)
- Collect data on receptive fields (Step 13)
- Engage in argument about alternative models (Steps 14–16)
- Review vocabulary and revisit the focus question (Steps 17–19)
- Read "Hearing" as homework and discuss next session (Steps 20, 21)

FOCUS QUESTION
How does the sense of touch work in humans?

SESSION 1 45–50 minutes

1. **Review organ-system interaction**
   Remind students that they have just spent a couple of weeks understanding how the organ systems in a human work together to sustain life. Pose these questions for students to discuss with a partner.

   ➤ *How do we as living organisms interact with the world around us?* [We look, hear, move, and so on.]

   ➤ *How do we know what is going on around us?* [We use our senses.]

   ➤ *What are the senses?* [Taste, touch, vision, hearing, and smell.]

   Tell students to think about what it would be like if we couldn't sense the environment around us. Ask one student in each group to imagine they do not have one of the following senses: sight, hearing, smell, taste, or touch.

   Students should quietly consider how the lack of the particular sense would affect their understanding and appreciation of the world. Ask them to write a short paragraph in their notebooks about what they imagine they would experience or miss.

   After a few minutes, ask students to share what they wrote with a partner or in their groups.

EL NOTE

Write the word "senses" and list them on the word wall.

For students who need a scaffold, provide a frame such as, "If I could not _____, then I would _____."

Human Systems Interactions Course—FOSS Next Generation **159**

INVESTIGATION 3 – The Nervous System

2. **Focus question: How does the sense of touch work in humans?**

   Tell students that in this investigation, they will find out more about how humans gather information about the world, how that information gets to the brain, and how the brain processes the information to guide actions. They will start off by exploring the sense of touch. Project or write the focus question on the board, and have students write it in their notebooks.

   ➤ *How does the sense of **touch** work in humans?*

   Give students a moment to record ideas about what kind of information about the world the sense of touch gives them. They should leave the rest of the page to return to at the end of this part.

> **EL NOTE**
>
> Have students make a word web for "touch" in their notebooks.

3. **Focus on the sense of touch**

   Ask,

   ➤ *How do you know something is touching you?* [You feel it.]

   Give groups a moment to consider what that means.

   ➤ *What do you mean by "feel it"?*

   If students struggle to answer this question, that's okay. Accept all answers for now.

   ➤ *What are the kinds of sensations you might experience when you touch something intentionally or when something touches you?* [Pressure, tickling, movement, pain, hot, cold, wetness, textures.]

   Things that we feel, either by reaching out to touch them intentionally, or by having something touch us, are part of our sense of touch. Let's explore the sense of touch a little more.

> **TEACHING NOTE**
>
> If students are curious about braille, you can tell them that it is a code based on a system of different configurations of six dots. It was developed in the early 1800s by Louis Braille (1809–1852), who became blind when he was 3. He published the first book in braille when he was 15 years old.

4. **Test fingertips and knuckles**

   Tell students that they will work in pairs to explore the sense of touch in their fingertips and knuckles. Show them the braille strips and explain that the strips contain a set of one, two, and three dots taken from the braille alphabet. Their job is to see if they can determine the number of dots, first using their fingertips and then using their knuckles.

   Distribute notebook sheet 7, *Touch*, and describe the procedure. Ask a student to help demonstrate, acting as the subject. You can take the part of the tester. Demonstrate, using a document camera if possible.

   - One student is the tester; the other is the subject.
   - The subject must keep his or her eyes closed during the entire test.

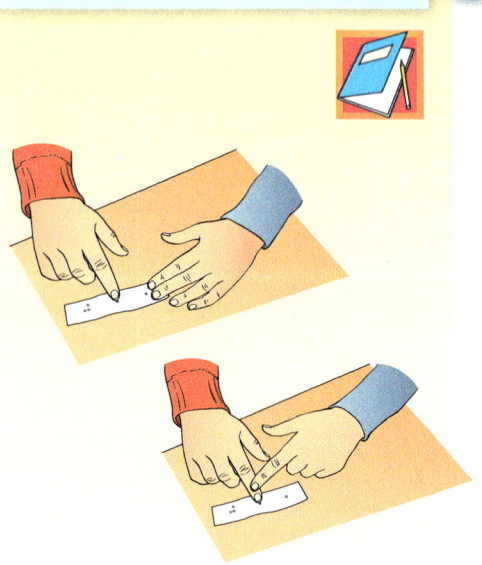

160   Full Option Science System

Part 1: The Sense of Touch

- The tester chooses the dot or dots that the subject will feel first and moves the strip under the subject's finger. The tester then gently pushes down the subject's finger so that the fingertip can feel the dot(s). (A light touch is usually more successful.)
- The tester records the number of dots on the subject's notebook sheet (not his or her own notebook sheet), and records if the subject had a correct response (Y or N). This continues for ten trials. The number of correct responses can be recorded in the final box of the table.
- Switch roles and repeat.
- When the fingertip test is done, test the knuckles in the same way. Use the first or second knuckle on any finger except the thumb.

5. ## Discuss results
   It shouldn't take long before students realize that they really can't tell how many dots they are feeling with their knuckles. Expect laughter and consternation. Tell students they don't need to answer the question on their notebook sheet yet. They will learn more about the sense of touch shortly.

   Ask,

   ➤ *Which is better for sensing braille dots, your fingertips or your knuckles?* [Fingertips.]

   ➤ *Why don't blind people use their knuckles to read braille?* [People can't feel things as easily with their knuckles. Knuckles are bonier.]

   Give students a minute to consider the following question in their groups and then solicit responses from the class.

   ➤ *Why do you think fingertips are more sensitive than knuckles?*

6. ## Introduce sensory receptors
   Tell students that anything that causes an action, or a **response**, in an organism is called a **stimulus**. When something presses on their skin, the stimulus is pressure. The skin has cells called **sensory receptors** that respond to pressure.

   Access the online "Touch Menu." Project the "Touch Receptors" activity and make sure that the sound is turned on. Click the Pressure button from the menu. This short demonstration explains how pressure receptors can be stimulated by touch. Tell students that pressure receptors are an example of receptors called **mechanoreceptors** because something mechanical, such as pressure, is the stimulus.

TEACHING NOTE

Some students will probably bring up the idea that they use their fingertips to sense things more than they use their knuckles, so they become more "sensitive." Do not disabuse them of that notion. After finding out about receptive fields, they will think about *why* they use their fingertips more.

EL NOTE

For Step 6, write these words on the word wall illustrating the "cause-and-effect" relationship, e.g., stimulus → response. Add the word *pressure* under *stimulus*. When introduced, add the word *mechanoreceptors*.

Point out the Greek origin of the suffix "mechano" meaning machine or mechanical and the Latin root word "receptor" meaning receive.

TEACHING NOTE

Students can look at other receptors and explore the "3D Finger" later. See the reference to the 3D finger resource in Step 10.

Human Systems Interactions Course—FOSS Next Generation

INVESTIGATION 3 – The Nervous System

SCIENCE AND ENGINEERING PRACTICES

Obtaining, evaluating, and communicating information

READING *in Science Resources*

7. **Read "Sensory Receptors"**
   Ask students to turn to the article "Sensory Receptors." Use the guide on the following pages to lead an in-class reading of the introduction and the "Mechanoreceptors" page of the article.

8. **Use a reading comprehension strategy**
   Explain that students will be using the sections in this article to learn more about the different types of sensory receptors. You might provide an example of a concept grid (see example below).

   Have students set up their notebooks to take notes. Start by reading aloud or having a volunteer read the title and the first sentence. Pause for students to discuss the question and their personal experiences with a partner.

   Tell students to read the rest of page 55 independently.

ELA CONNECTION

This suggested strategy addresses the Common Core State Standards for ELA for literacy.

SL 1: Engage in collaborative discussions.

Sensory Receptor	Function	Stimuli	Location	Example
Mechanoreceptors				
Chemoreceptors				
Photoreceptors (Electromagnetic)				

TEACHING NOTE

For more information on scaffolding literacy in FOSS, see the Science-Centered Language Development in Middle School chapter in *Teacher Resources*.

162   Full Option Science System

Part 1: The Sense of Touch

Sensory Receptors

Mmm, I smell doughnuts! I can picture which one I want before I even see them. Why does that smell bring up images and memories in my brain?

One of the jobs of your nervous system is to take in information, decide what it means, and respond to it in some way.

For many, the smell of doughnuts means something tasty. The brain compares the smell to prior experiences. The smell stimulates a memory of a doughnut, a warm, chocolate treat. The brain recognizes desire and responds by sending the body to find the doughnut and eat it.

Our senses all rely on the ability of **sensory receptors** to take in information and translate it into electric impulses. These impulses travel to the brain. Each sense has receptors that respond to certain inputs: mechanical, chemical, or **electromagnetic**.

We experience the world around us through our senses, which work together to collect information and send it to the brain. How might all five senses help you enjoy a doughnut?

Investigation 3: The Nervous System 55

A Determine the central idea: When students finish reading the page, have them write one sentence, make a simple drawing or use a graphic organizer to capture the important idea of this passage.

B Discuss collaboratively: Give students a few moments to share their thoughts about this question with a partner.

C Read and comprehend: Explain that they will continue to read the section about mechanical receptors now and will read about the other types—chemical and electromagnetic—later in the investigation.

ELA CONNECTION

These suggested strategies address the Common Core State Standards for ELA for literacy.

RST 2: Determine the central ideas or conclusions of a text; provide an accurate summary.

RST 10: Read and comprehend science/technical texts independently and proficiently.

SL 1: Engage in collaborative discussions.

Human Systems Interactions Course—FOSS Next Generation 163

INVESTIGATION 3 – *The Nervous System*

Gather relevant information: Refer to the concept grid or have students make a section entitled, "mechanoreceptors" to take notes about this type of sensory receptor. They should include the function, stimuli, location, and an example.

Think question: Give students a few moments to share their thoughts about this question with a partner.

Save the last three pages of this reading for Part 3 of this investigation.

ELA CONNECTION

These suggested strategies address the Common Core State Standards for ELA for literacy.

WHST 8: Gather relevant information from multiple print and digital sources.

SL 1: Engage in collaborative discussions.

Mechanoreceptors

You investigated mechanical inputs in the **sense of touch**. Your skin has several types of pressure receptors that respond to touch. Each **mechanoreceptor** sent a message to your brain to tell it about the touch.

Other mechanoreceptors help you get to the bathroom in time. When your bladder is about half full, stretch receptors send signals along your pelvic nerves to your spinal cord.

A signal returns to muscles in your bladder, telling certain muscles to contract. Those contractions put pressure on the bladder. That pressure makes you want to urinate.

Think Question

Where else might mechanoreceptors be inside the body?

164

Full Option Science System

Part 1: The Sense of Touch

9. Read "Touch"

Tell students to turn to the article titled "Touch" on page 60 to find out more about touch receptors. Read aloud or have a volunteer read the title and first paragraph. Pause for students to discuss why they think the sense of touch is so important. Encourage them to use evidence and reasoning to support their claims.

10. Use a reading comprehension strategy

Explain that students will continue to gather information from the readings to learn more about sensory receptors. They should continue to record information as they read by using the concept grid they set up for the previous reading. They could also make their own diagrams in their notebooks modeled after the diagrams in the article. For example, they could make a diagram of the skin showing the different kinds of receptors. Students could also view the online activity "3D Finger" to confirm the labels on their diagram of the skin.

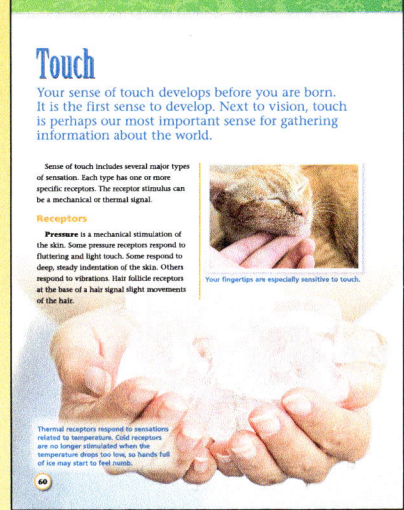

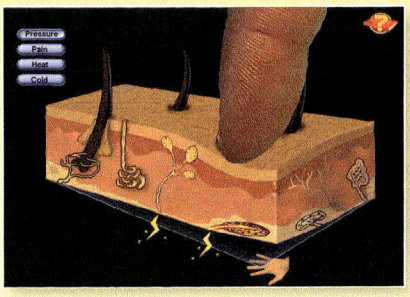

SCIENCE AND ENGINEERING PRACTICES

Obtaining, evaluating, and communicating information

TEACHING NOTE

For more information on scaffolding literacy in FOSS, see the Science-Centered Language Development in Middle School chapter in Teacher Resources.

Human Systems Interactions Course—FOSS Next Generation

INVESTIGATION 3 – The Nervous System

Gather relevant information: Continue reading the next paragraph and pause to model how students might use a tree map to organize new information about touch. See the example of the tree map at the bottom of this page.

Continue to gather information: Tell students to continue reading this page and the next and to record new information on their concept grids, skin drawings, or graphic organizers.

ELA CONNECTION

This suggested strategy addresses the Common Core State Standards for ELA for literacy.

WHST 8: Gather relevant information from multiple print and digital sources.

Touch

Your sense of touch develops before you are born. It is the first sense to develop. Next to vision, touch is perhaps our most important sense for gathering information about the world.

Sense of touch includes several major types of sensation. Each type has one or more specific receptors. The receptor stimulus can be a mechanical or thermal signal. A

Receptors

Pressure is a mechanical stimulation of the skin. Some pressure receptors respond to fluttering and light touch. Some respond to deep, steady indentation of the skin. Others respond to vibrations. Hair follicle receptors at the base of a hair signal slight movements of the hair.

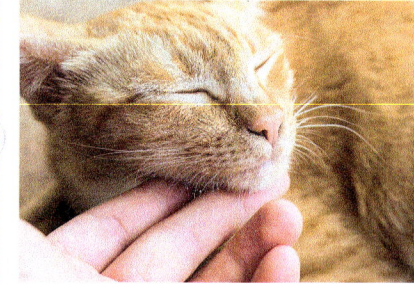

Your fingertips are especially sensitive to touch.

Thermal receptors respond to sensations related to temperature. Cold receptors are no longer stimulated when the temperature drops too low, so hands full of ice may start to feel numb.

60

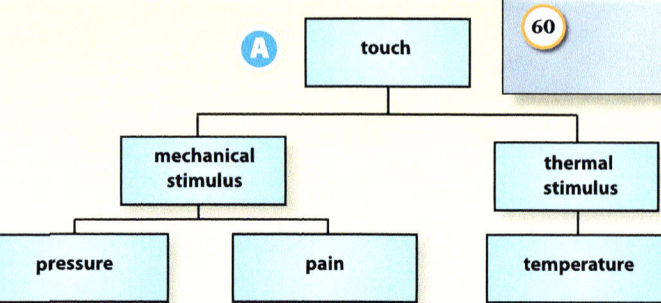

Full Option Science System

Part 1: The Sense of Touch

Pain and temperature signals are sent by bare nerve endings embedded in the skin. Some receptors sense hot or cold. Other bare-ending receptors sense sharp pain, deep aches, or damage caused by burning or freezing of skin.

Look at the diagram. Each receptor sends a message through nerves to the spinal cord. From there, the message travels to the brain. The brain processes the touch received on that part of the skin.

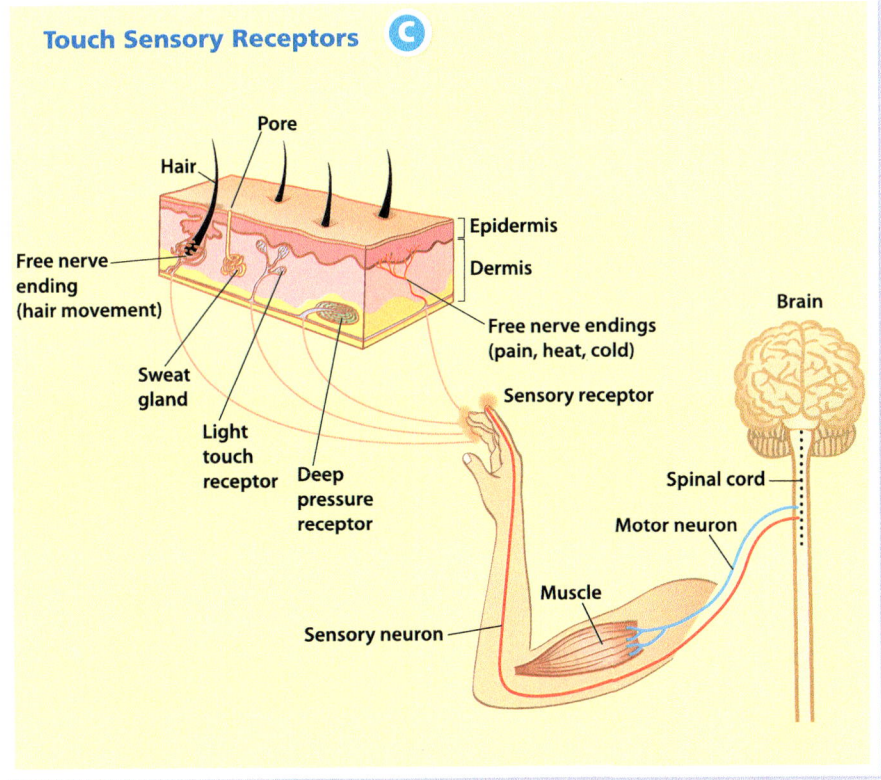

Touch receptors in the skin are specialized to detect different kinds of touch. Some respond to temperature, others to pain and to pressure, and still others to the textures of what we touch.

Investigation 3: The Nervous System 61

Integrate technical information: Give students a few moments to look at the diagram. Remind them of questions they can ask themselves to help them with comprehension, such as:

➤ *What is the purpose of this diagram?*

➤ *What parts do I recognize? Which parts are new to me?*

ELA CONNECTION

This suggested strategy addresses the Common Core State Standards for ELA for literacy.

RST 7: Integrate quantitative or technical information expressed in words in a text with a version of that information expressed visually.

Human Systems Interactions Course—FOSS Next Generation **167**

INVESTIGATION 3 – The Nervous System

Read and comprehend science: Have students read the rest of the article independently. Remind them to analyze the goose bumps diagram before and after reading.

Discuss collaboratively: Give students a moment to share something new they learned from the article and a personal connection.

Touch is as important to a baby's health and development as eating and sleeping.

Sensory Facts

- Being touched is important for normal human development. If babies are not cuddled a lot, they grow slowly, have trouble learning, and suffer physically. They also do not get emotionally close to other people.

- Many sensations result from combined signals to the brain. For example, no touch receptors sense wetness. Cold receptors and pressure receptors both send signals. The brain combines them to sense that the skin is wet.

- Goose bumps are the result of signals from cold receptors in the skin. The brain responds to cold with a message to the blood vessels and smooth muscles in the skin. The result is small bumps that help reduce heat loss.

- Itching is caused by an irritation to the skin. Irritants can be infections, allergies, insects, and fabrics. The irritant moves small touch receptors in the skin, which produces itching. Scratching can relieve the irritation. If scratching causes pain receptors to fire, that neutralizes the itching sensation.

ELA CONNECTION

This suggested strategy addresses the Common Core State Standards for ELA for literacy.

RST 10: Read and comprehend science/technical texts independently and proficiently.

SL 1: Engage in collaborative discussions.

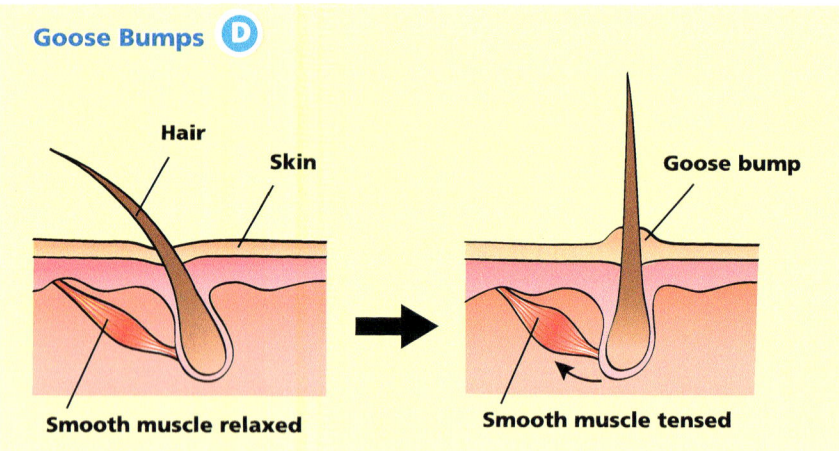

Goose Bumps

Goose bumps and hair "standing on end" are automatic responses to cold or strong emotions like fear. Tiny muscles around hair follicles contract, bunching up the skin and pushing up the hairs.

62

168   Full Option Science System

Part 1: The Sense of Touch

- Say you have a sharp pain in a small area, such as a pinprick. Rubbing the area relieves pain. Rubbing activates pressure receptors in the skin. This decreases the intensity of the pain receptors' message to the brain.
- In the disease of leprosy, pain receptors are damaged. They do not send pain messages when parts of the body are hurt. As a result, people with leprosy do not know when they are harming their bodies. This lack of awareness can lead to a lot of damage, especially to hands and feet. Parts of the body that are damaged can become infected and inflamed. They can decay if untreated.
- The star-nosed mole has a great sense of touch. Its nose tentacles have six times more touch receptors than a human hand. As the mole tunnels in the dirt, its tentacles can touch 10 to 12 objects per second.

Think Questions

1. Why don't you feel your watch, earrings, underwear, or socks after you put them on?
2. What do you think a web-building spider uses more to capture prey, its sight or sense of touch? Why?
3. What is the most useful thing you discovered about touch?

The supersensitive nose of the star-nosed mole has a ring of 22 short, pink tentacles. They help it instantly identify prey by touch.

Investigation 3: *The Nervous System* 63

Discuss the reading: Give students a few quiet moments to jot down their responses to the Think Questions and then have them discuss their thoughts with their groups. Encourage them to use evidence from the text to support their ideas.

▶ *Why don't you feel your watch, earrings, underwear, or socks after you put them on?* [Sensory adaptation occurs when sensory receptors decrease their sensitivity to stimuli. Sensory adaptation is one way the brain protects itself from overload.]

▶ *What do you think a web-building spider uses more to capture prey, its sight or sense of touch? Why?* [Sense of touch; the web provides a wider area for the spider to observe by touch, that is, to sense vibrations from moving prey.]

▶ *What is the most useful thing you discovered about touch?*

ELA CONNECTION

This suggested strategy addresses the Common Core State Standards for ELA for literacy.

SL 1: Engage in collaborative discussions.

Human Systems Interactions Course—FOSS Next Generation **169**

INVESTIGATION 3 – The Nervous System

SESSION 2 45–50 minutes

11. Discuss pressure sensory receptors

Give students a moment to remember what they experienced in the previous session and think about the pressure receptors in their fingertips and knuckles. Ask,

➤ *Why are your fingertips better able to discriminate the number of small dots in a braille code?* [Students might suggest one of the following: The pressure receptors in the fingertips are more sensitive. There are more receptors in the fingertips.]

Record students' responses on the board, making sure to include the misconception that the receptors in the fingertips are more sensitive. Then tell students,

What if I told you that pressure receptors all work in the same way. They are all equally sensitive to touch. With that new information, how might you revise your model of what is different between the receptors in your fingertips and knuckles? Discuss your ideas with a partner.

After students have shared ideas with a partner, focus on the suggestion that there must be more pressure receptors in the fingertips than there are in the knuckles.

12. Introduce models for pressure receptive fields

Project teacher master C, *Pressure Receptive-Fields Models A* using a document camera. Tell students,

Each sensory receptor cell has a **receptive field**, *an area from which it gathers information. Here are two models for pressure receptive fields.*

➤ *How are these two models different?* [Size of the receptive field, either small or large; in both cases, the fields overlap.]

Provide more information about the models.

Each receptor cell can send only one message to the brain no matter how many points in its field are stimulated.

Demonstrate this by touching one of the large receptive fields in the top image with one finger. Ask,

➤ *How many messages are sent to the brain?* [One.]

Now touch the same receptive field with two fingers. Ask,

➤ *How many messages are sent to the brain?* [One.]

SCIENCE AND ENGINEERING PRACTICES

Developing and using models

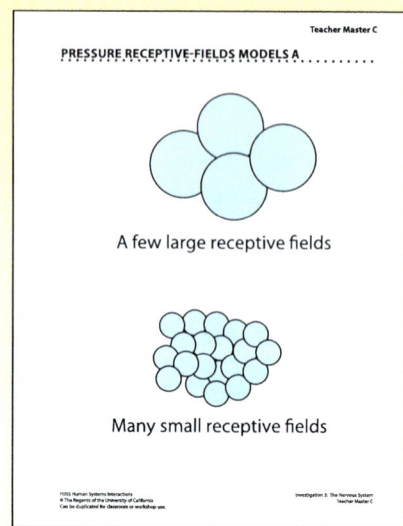

Teacher Master C

170 Full Option Science System

Part 1: The Sense of Touch

Move your fingers apart and touch two different receptive fields with two fingers. Ask,

➤ *How many messages are sent to the brain?* [Two.]

Demonstrate the same principle using the smaller receptive fields.

Remind students that they told you that the fingertips may have more sensory receptors than the knuckles.

13. Collect data on receptive fields

Ask students how the class might find evidence to determine if the fingertips and knuckles have a few large receptive fields or many small receptive fields. Give students a few minutes to discuss their ideas with a partner. Call on a few volunteers to share their idea or what their partner said.

Show students a large paper clip and suggest that they could use it as a tool to test their pressure receptors. Unfold it and move the ends until they are 5 mm apart. Discuss a procedure for using the paper clip to test pressure points.

The tester will touch the paper-clip points several times to the knuckle of the subject, sometimes touching just one point and sometimes both, to see if the subject can tell the difference. Subjects close their eyes and try to sense how many points they feel. Testers then test the fingertip in the same way. Then students switch roles.

Have Getters get two paper clips for their group, and let the testing begin. This should take no more than a minute or two. When students are done, ask,

➤ *What was the difference between feeling the points with your fingertips and with your knuckles?* [I could always feel the difference between one point and two points on my fingertip and usually not on my knuckle.]

➤ *What does that tell you about the receptive fields and the sensory receptors in the fingertips?* [There could be smaller and more numerous receptive fields in the fingertips so there could be more receptors.]

Confirm that research has shown that there are more pressure receptors in the fingertips. These sensory receptors are small and packed close together, enabling the fingers to discriminate between very small differences in the texture of surfaces, such as the number of braille dots on a piece of paper.

TEACHING NOTE

In the skin, the receptive fields may overlap even more than they do on the teacher master. This is a limitation of this model.

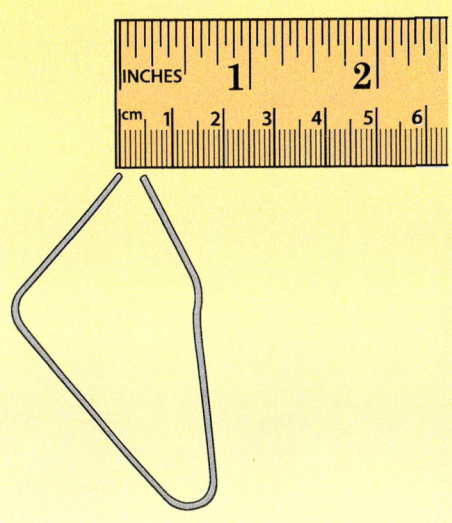

SCIENCE AND ENGINEERING PRACTICES

Planning and carrying out investigations

Analyzing and interpreting data

EL NOTE

Suggest students make a diagram in their notebooks to show how they think the receptive fields might look in the fingertips compared to those in the knuckles.

Human Systems Interactions Course—FOSS Next Generation 171

INVESTIGATION 3 – *The Nervous System*

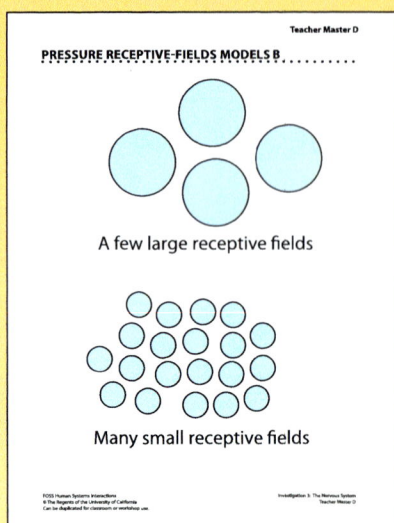

Teacher Master D

14. Display alternative models

Project teacher master D, *Pressure Receptive-Fields Models B*. Ask students how these models for the pressure receptive fields are the same and different from the previous models. They should notice that these fields are large and small but they do not overlap.

Ask students how they could gather data to determine which models (overlapping or nonoverlapping) better describe the pressure receptors and receptive fields on their hands. Have them discuss this in their groups.

15. Engage in argument from evidence
Ask

➤ *What data could you collect to develop evidence to support the overlapping model or the nonoverlapping model?*

Let groups discuss what they could do and then give them a few minutes to collect the data.

Circulate among the groups and see how they approach the problem. They should be using the paper clip testing technique again, and searching for areas on their fingertips and knuckles where they could not sense pressure. That would provide evidence for nonoverlapping. If they can always sense pressure, that would provide evidence for overlapping.

16. Assessing progress: notebook entry

Ask students to select which model they can support with evidence and to describe that evidence. Ask them to write in their notebook 1) the test; 2) the data they gathered; and 3) their claim and their argument to support that claim. They should include information on both the size of the receptive field and the overlapping aspect.

What to Look For

- *Students describe a reasonable and appropriate paper-clip test.*
- *Students clearly describe the data they collected from their test.*
- *Students use the data to develop evidence to construct an explanation to support one of the models, such as, if they always sensed the paper clip, that supports the overlapping model; or if there were times they did not sense the paper clip, that supports the non-overlapping model.*

SCIENCE AND ENGINEERING PRACTICES

Developing and using models

Constructing explanations

Engaging in argument from evidence

Full Option Science System

Part 1: The Sense of Touch

17. Introduce the homunculus

Project teacher master E, *Homunculus*. Say,

Scientists represent the number of touch receptors in different parts of the body using a diagram called a homunculus (huh•MUN•cue•less), which is Latin for "little person." Talk to your partner about what you notice about this image. Focus on the structure and function of each body part portrayed here and why some are exaggerated.

Confirm that the image exaggerates different parts of the body to show how touch sensitive they are. Then hold a class discussion using the following questions. Students can also use the concept of scale and proportion to interpret and explain the diagram.

➤ *Which parts of the body are best equipped to gather information through the sense of touch and which are poorly equipped?* [The lips, tongue, hands, and feet are best; the ears, legs, and arms are poorly equipped.]

➤ *Why might it be important to have certain parts of the body be very sensitive to touch and other parts not?* [We use certain parts of our body to interact with the world. Those parts should have more sensory receptors.]

18. Review vocabulary

Give students a few moments to review the vocabulary developed in this part. Have them update their vocabulary indexes and tables of contents if they haven't already done so.

19. Answer the focus question

Tell students to write a response to the focus question. They should include what they have learned about sensory receptors and receptive fields.

➤ *How does the sense of touch work in humans?*

Once students have completed their own responses, ask a few to share their thoughts with the class.

TEACHING NOTE

This is a structure/function relationship. Structures such as fingers, lips, and feet (which help us balance) transmit important information to the brain. The skin on your back does not provide as much important sensory information that you need to respond to. Therefore, it is not important that your back be as sensitive to touch as your fingertips are.

CROSSCUTTING CONCEPTS

Scale, proportion, and quantity

Structure and function

mechanoreceptor
receptive field
response
sensory receptor
stimulus
touch

EL NOTE

For students who need scaffolding, provide sentence frames such as "Humans have different kinds of touch receptors to gather information from the environment. For example _____. We found out that _____."

Human Systems Interactions Course—FOSS Next Generation

INVESTIGATION 3 – The Nervous System

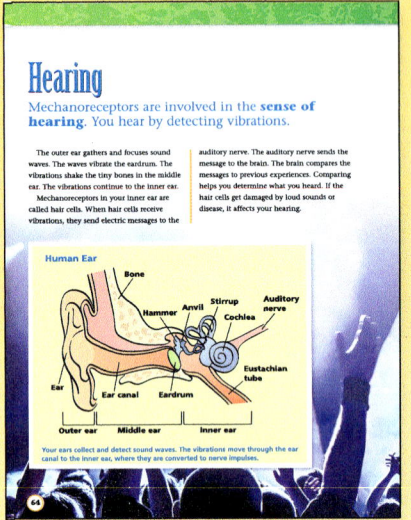

ELA CONNECTION

These suggested strategies address the Common Core State Standards for ELA for literacy.

RST 10: Read and comprehend science/technical texts independently and proficiently.

WHST 9: Draw evidence from informational texts to support analysis, reflection, and research.

EL NOTE

Make sure that students understand that "vibrate" means to move back and forth.

20. **Extend the investigation with homework**
    Have students read the article "Hearing" in *FOSS Science Resources* as homework and do the following.

    a. Analyze the human ear diagram before and after reading. Use the diagram to explain to a friend or relative the process of hearing.

    b. Add new information to your notes and/or Sensory Receptors concept grid.

    c. Answer the Think Questions in your notebook and be prepared to discuss them with your group the next day. Remember to cite evidence from the text to support your ideas.

WARM-UP

21. **Discuss reading about hearing**
    The next day, spend a few minutes discussing the answers to the first question as the warm-up for Part 2.

    ➤ *How are hearing and touch alike and how are they different?* [Both touch and hearing involve mechanoreceptors that send messages to the brain. The skin has pressure receptors that respond to touch. The ear has hair cells that respond to the vibration of sound waves.]

Part 2: Sending a Message

MATERIALS for
Part 2: Sending a Message

Provided equipment

For each student
- 1 *FOSS Science Resources: Human Systems Interactions*
- "Brain Messages"
- "Neurotransmission: The Body's Amazing Network"
- "Brain Map of Sensory Activity"

For the class
- 28 Plastic cups
- 4 Zip bags
- 28 Aluminum-foil balls (see Getting Ready)
- String
- 4 Timers accurate to 0.01 second
- 1 Hole punch

Teacher-supplied items

For each group (of 8 students)
- 1 Mini-whiteboard, pen, and eraser (optional)

For the class
- 4 Sets of Neural-Message Relay cards (see Getting Ready)
- 12 Pieces of card stock, 22 × 28 cm (8.5" × 11") (optional)
- Self-stick notes (for marking reading)

FOSSweb resources

For each student
- Notebook sheet 8, *Neural Pathways*
- Notebook sheet 9, *Response Sheet—Investigation 3*
- Teacher master F, *Sending the Message* (optional)

For the class
- Online activity: "Brain: Synapse Function"
- Online activity: "Brain: Neuron Growth"

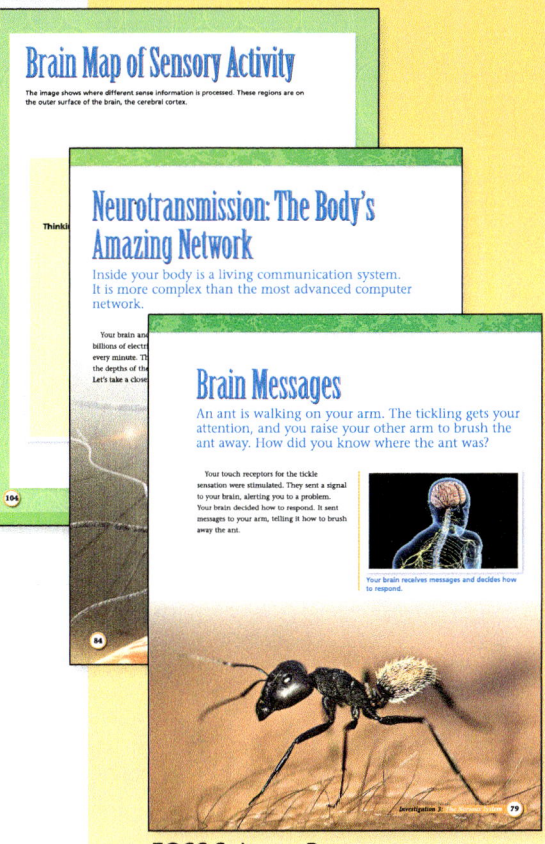

FOSS Science Resources

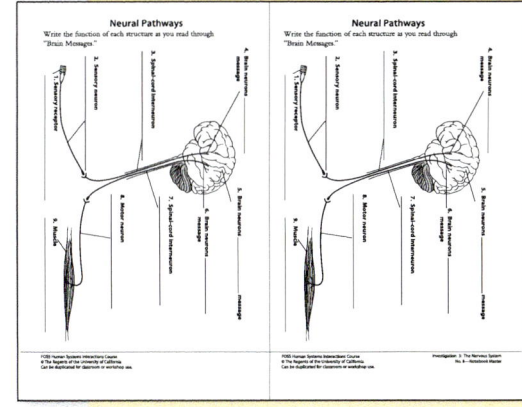

No. 8—Notebook Master

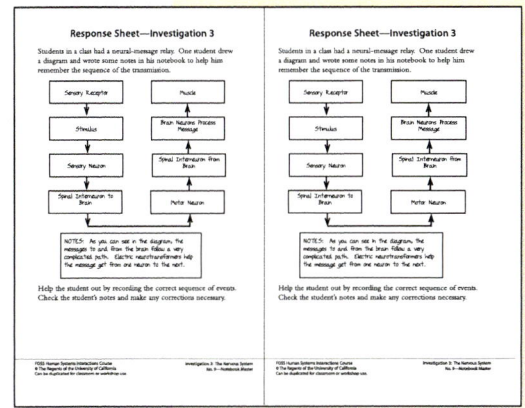

No. 9—Notebook Master

Human Systems Interactions Course—FOSS Next Generation   175

INVESTIGATION 3 – *The Nervous System*

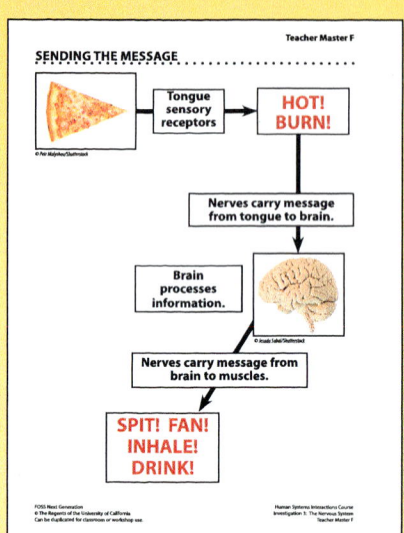

Teacher Master F

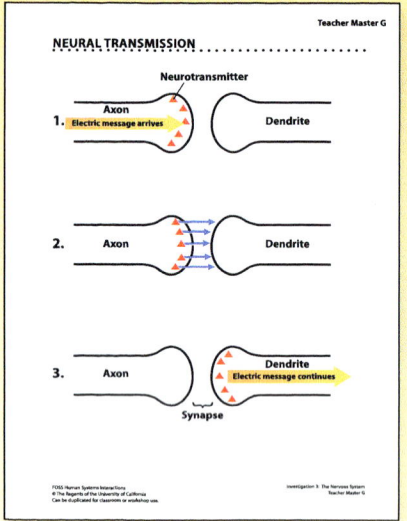

Teacher Master G

For the teacher

- Teacher master F, *Sending the Message*
- Teacher master G *Neural Transmission*
- Teacher master H–J, *Neural-Message Relay Cards A–C*
- Teacher master K, *Neural-Message Relay Setup*
- *Embedded Assessment Notes*
- Teaching slides, 3.2

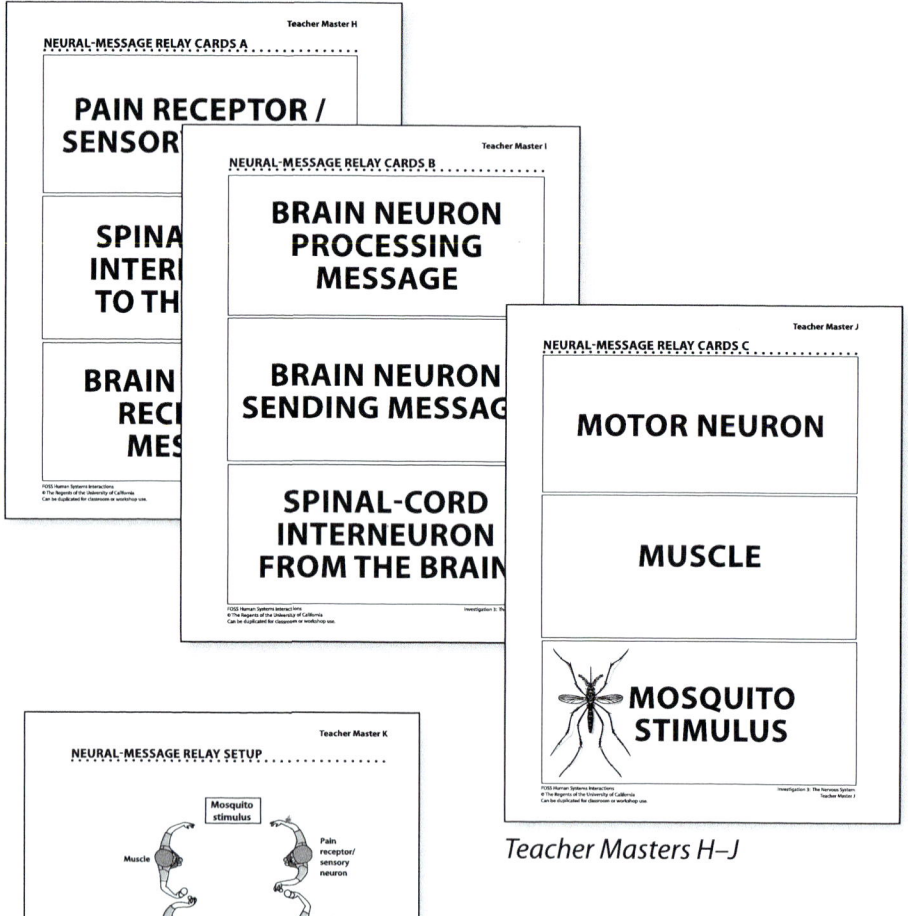

Teacher Masters H–J

Teacher Master K

176 Full Option Science System

Part 2: Sending a Message

GETTING READY for
Part 2: Sending a Message

Quick Start

Schedule	3 active investigation sessions, including two readings
Preview	• Preview the FOSSweb Resources by Investigation for this part (such as printable masters, teaching slides, and multimedia) • Preview the image: "Brain Map of Sensory Activity," Step 1 • Preview the reading: "Brain Messages," Step 5 • Preview online activities: "Brain: Neuron Growth," Step 15 "Brain: Synapse Function," Step 16 • Preview the reading: "Neurotransmission: The Body's Amazing Network," Step 18
Print or Copy	**For each student** • Notebook sheets 8, 9 • Teacher master F (optional) **For each group of eight** • Teacher masters H–J **For the teacher** • *Embedded Assessment Notes*
Prepare Material	• Prepare simulated neurotransmitters **A** • Prepare neural-message relay cards **B** • Prepare bags of group materials **C**
Plan for Assessment	• Review Step 17, "What to Look For" in the response sheet

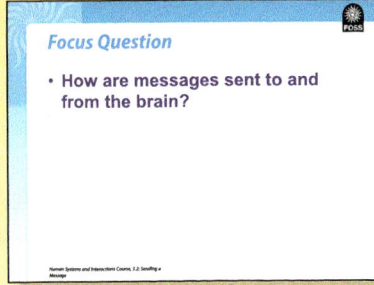

Teaching slides, 3.2

Human Systems Interactions Course—FOSS Next Generation

INVESTIGATION 3 – The Nervous System

> **NOTE**
> Students will work in large groups of eight for the neural-message relay. This means you will need four sets of materials for most classes.

Preparation Details

A ▶ Prepare simulated neurotransmitters

The first time you do the simulation, you will need to make several aluminum-foil balls for simulated neurotransmitters. Each group of eight students needs seven balls.

Tear about 30 cm of aluminum foil off the roll. Crumple the foil into a ball about the size of a table-tennis ball. You might want to make some extras in case any get lost.

B ▶ Prepare neural-message relay cards

Each group of eight students will need a set of the nine neural-message relay cards. Make copies of teacher masters H–J for the groups. You may wish to copy the teacher masters onto card stock so you can use the cards in future years.

Cut out the cards and punch holes near the top of each one. Tie a length of string to each so students can wear the cards like necklaces. Do not put string on the mosquito-stimulus card, however, because the card will either rest on the table or be in the right hand of the student who plays the pain receptor/sensory neuron.

C ▶ Prepare bags of group materials

Put seven plastic cups, seven aluminum-foil balls, and nine neural-message relay cards in a large zip bag for each group. You will use the sets of materials throughout the day.

Part 2: Sending a Message

GUIDING *the Investigation*
Part 2: *Sending a Message*

SESSION 1

Students will . . .
- Discuss receptors and the "Brain Map of Sensory Activity" (Step 1)
- Analyze an example of stimulus and response (Steps 2–4)
- Read and discuss "Brain Messages" (Steps 5, 6)
- Preview a neural-message relay in terms of synapse and neurotransmitters (Steps 5, 6)

SESSION 2

Students will . . .
- Take part in a neural-message relay simulation (Steps 9–13)
- Analyze the relay in terms of timing and practice (Step 14)
- Discuss neuron growth in humans (Step 15)
- View online activity "Brain Synapse Function" (Step 16)
- Communicate understanding on response sheet (Step 17)

SESSION 3

Students will . . .
- Read and discuss "Neurotransmission: The Body's Amazing Network" (Steps 18, 19)
- Review vocabulary (Step 20)
- Answer the focus question (Step 21)
- Revisit the response sheet (Step 22)

FOCUS QUESTION
How do messages travel to and from the brain?

SESSION 1   45–50 minutes

1. **Focus on the brain**
   Review the sensory receptors in the skin. Ask,

   ➤ *What kinds of sensory receptors are in the skin?* [Mechanoreceptors called pressure receptors, pain receptors, and temperature receptors.]

   ➤ *How does your brain know the difference between a pain stimulus, a pressure stimulus, and a temperature stimulus?* [Students may not know.]

   Students may say that different receptors are activated by different kinds of stimuli, different parts of the brain respond to different stimuli, there are different kinds of signals, and so on.

EL NOTE
Students can use their notebook entries to review the labeled diagram of touch receptors.

Human Systems Interactions Course—FOSS Next Generation   **179**

INVESTIGATION 3 – The Nervous System

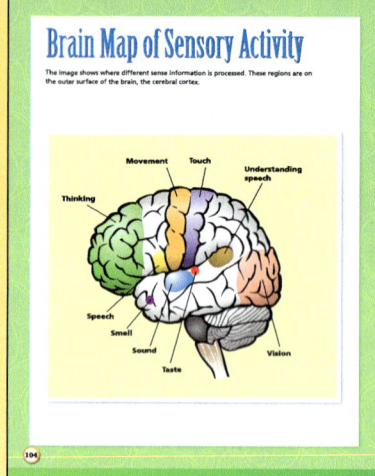

SCIENCE AND ENGINEERING PRACTICES
Constructing explanations

CROSSCUTTING CONCEPTS
Cause and effect

EL NOTE
Use a cause-and-effect organizer as you record student responses to help them think about the relationship between stimulus and response.

Have students turn to the "Brain Map of Sensory Activity" image in *FOSS Science Resources*. Look at the different parts of the brain and discuss what you notice with a partner. Ask,

▶ *What do you notice?* [Different senses are processed in different parts of the brain.]

Tell students,

The outer folded layer of the brain is where most of the sensory processing happens. The outer layer is called the **cerebral cortex**.

2. **Respond to a stimulus**

Ask students to imagine a hot pizza right out of the oven.

▶ *If you grab a slice of that pizza and sense that it's warm and put it in your mouth only to find that the pizza is extremely hot, what do you do? Tell your partner what you would do. How would you respond?*

Listen for and confirm answers that relate to trying to cool off the burned tongue or mouth. Then pose this question for students to discuss.

▶ *What do the sensory receptors in your mouth do?* [Send a message to the brain that the mouth is burning.]

Ask students to explain this process in terms of cause and effect.

▶ *What is the stimulus? What is the response?* [The stimulus in our example is a burning hot pizza. We respond by trying to cool off our mouth.]

Give groups a minute to think about the following question.

▶ *How do you know how to respond?*

Call on a student or two to share the group's ideas. Here are the points to look for:

- The brain receives a message from the mouth or tongue.
- The brain has to decide if it's dangerous or just uncomfortable.
- The brain somehow tells the mouth to spit out the hot food, or a hand to immediately start fanning in front of the mouth, or lungs to inhale, or to take a drink of something cold.

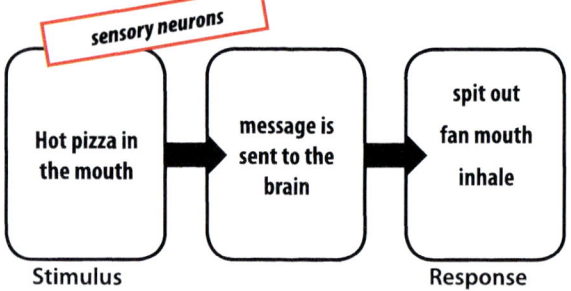

180

Full Option Science System

Part 2: Sending a Message

3. **Focus question: How do messages travel to and from the brain?**

   Tell students that the focus here is on finding the pathway between the sensory receptors and the brain, and the pathway between the brain and the part of the body that needs to respond. Pose the focus question, write or project it on the board, and have students write it in their notebooks.

   ➤ *How do messages travel to and from the brain?*

   Students do not need to answer the question at this time. They should leave space under the question to return to it later.

4. **Send the message**

   Project teacher master F, *Sending the Message*. Point to each part of the flow chart as you describe the story. Tell students,

   *When the hot pizza touches the tongue, the heat receptors in the tongue send a message in the form of an electric impulse along **nerves** to the brain, where the information is quickly processed. The brain knows that the tongue is in danger of being burned and quickly forms a response message to the muscles in the arm to fan the hand in front of the mouth.*

   The response message is sent out from the brain, through the spinal cord, and along the nerves until it reaches the mouth and arm muscles. There the nerve endings pass the message to the muscles to contract and move the tongue or hand.

   *Nerves are the "electric wires" that make up the brain and the structures of the nervous system. Nerves carry information in the form of electric impulses between the brain and the rest of the body. Each nerve is made of specialized cells called **neurons**.*

   Write "nerve" and "neuron" on the board to help students start to differentiate between them. Nerves are made of many cells called neurons.

> **TEACHING NOTE**
>
> *In case of danger (such as a burning mouth), the neural message would most likely be sent through a short local pathway called a reflex arc, speeding up the response, so that tissue is not damaged. For the sake of our story, however, we will use the brain messaging system, which is milliseconds slower.*

Human Systems Interactions Course—FOSS Next Generation   **181**

INVESTIGATION 3 – The Nervous System

SCIENCE AND ENGINEERING PRACTICES

Obtaining, evaluating, and communicating information

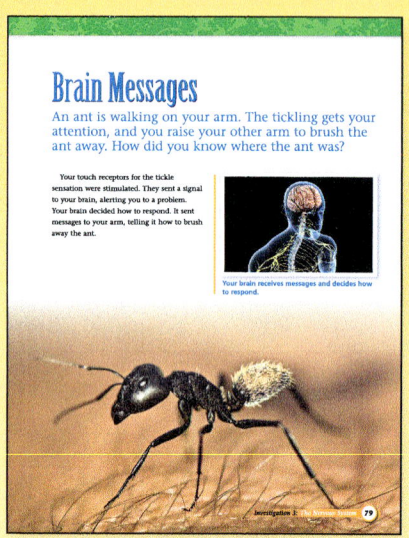

ELA CONNECTION

These suggested strategies address the Common Core State Standards for ELA for literacy.

RST 2: Determine the central ideas or conclusions of a text; provide an accurate summary.

RST 10: Read and comprehend science/technical texts independently and proficiently.

L 4: Determine or clarify meaning of unknown words or phrases.

TEACHING NOTE

For more information on scaffolding literacy in FOSS, see the Science-Centered Language Development in Middle School chapter in Teacher Resources.

READING in Science Resources

5. Read "Brain Messages"

Give students a few minutes to discuss the flowchart on teacher master F, *Sending the Message*. Have them copy it in their notebooks (or give them a copy of the teacher master) and write three questions they have next to any of the steps in the flow.

Tell them that "Brain Messages" will give them more information and help answer their questions. Use the guide on the following pages to conduct an in-class reading of the article.

Tell students to close their eyes and "feel" the text as you or a volunteer read aloud the first paragraph. Then, have students open their eyes and give them a few moments to discuss the question. Encourage them to use the ant photograph to describe their thoughts. Have them read the rest of the page with a partner and discuss what they know and what questions they have about the text so far.

6. Use a reading comprehension strategy

Have students read and discuss the rest of the article in their groups. You might want to assign roles to help keep students on track (Facilitator, Researcher, Recorder, Reporter). Students will use notebook sheet 8, *Neural Pathways*, during this session. Explain the process as follows:

1. Read the article independently. Use self-stick notes to mark unknown words and phrases, confusing text, and important ideas. Look for answers to the questions you noted on your *Sending the Message* flowcharts.

2. **Facilitator** leads a share-out of notes and asks the **Researcher** to determine or clarify the meaning of unknown words or phrases for the group.

3. **Facilitator** assigns each member a section to reread and briefly summarizes for the rest of the group.

4. The **Recorder** distributes notebook sheet 8, *Neural Pathways*, to each student and leads the rest of the group in discussing and writing in the function for each structure.

5. The **Reporter** shares out during the whole-class discussion what the group discussed and recorded on the sheets and any questions that are still lingering.

182   Full Option Science System

Part 2: Sending a Message

Brain Messages

An ant is walking on your arm. The tickling gets your attention, and you raise your other arm to brush the ant away. How did you know where the ant was?

Your touch receptors for the tickle sensation were stimulated. They sent a signal to your brain, alerting you to a problem. Your brain decided how to respond. It sent messages to your arm, telling it how to brush away the ant.

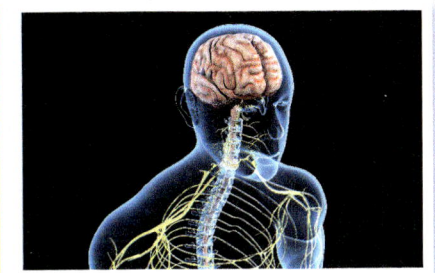

Your brain receives messages and decides how to respond.

A

Discuss purpose: Have students share any thoughts or personal connections they have about the introduction. Then ask,

➤ *Why do you think the author used this example? Can you think of other examples?*

➤ *How do the images help explain the process described in the text?*

ELA CONNECTION

This suggested strategy addresses the Common Core State Standards for ELA for literacy.

RST 6: Analyze the author's purpose in providing an explanation, describing a procedure, or discussing an experiment in a text.

Investigation 3: The Nervous System 79

Human Systems Interactions Course—FOSS Next Generation

INVESTIGATION 3 – The Nervous System

B

Read aloud or call on a volunteer to reread this section. Have students pause to consider "several hundreds of billions of neurons." Ask

➤ *Why do you think we have so many neurons?*

Give students a minute to locate the three basic parts of a neuron and discuss their structure and function.

➤ *How do you think the shapes of the different parts relate to their function?*

CROSSCUTTING CONCEPTS

Structure and function

ELA CONNECTION

This suggested strategy addresses the Common Core State Standards for ELA for literacy.

RST 7: Integrate technical information expressed in words in a text with a version of that information expressed visually.

B

Neurons at Work

The cells that make up your brain and all your nerves are called **neurons**. You have several hundreds of billions of neurons throughout your body and brain. They constantly send messages from one place to another.

Every neuron has three basic parts: a **cell body**, **dendrites**, and an **axon**. The cell body contains the nucleus of the cell. Extending from the cell body are numerous branches called dendrites (dendro = tree). Incoming electric messages enter the dendrites and travel to the cell body. Then they leave the cell body on a long, thin extension of the cell called the axon. The message passes from the axon to the dendrite(s) of one or more neurons. They carry the message on its way.

Two Interacting Neurons

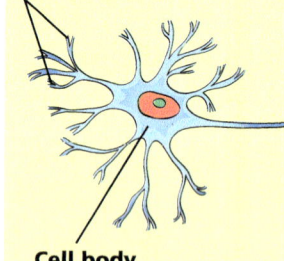
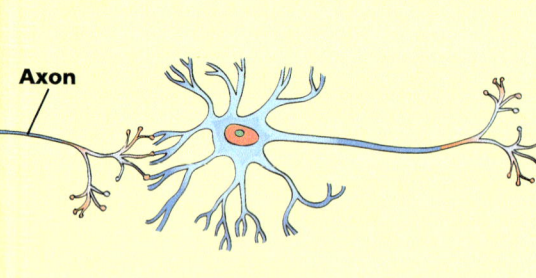

A message is picked up by dendrites, travels through the axon, and passes to the next neuron.

80

184 Full Option Science System

Part 2: Sending a Message

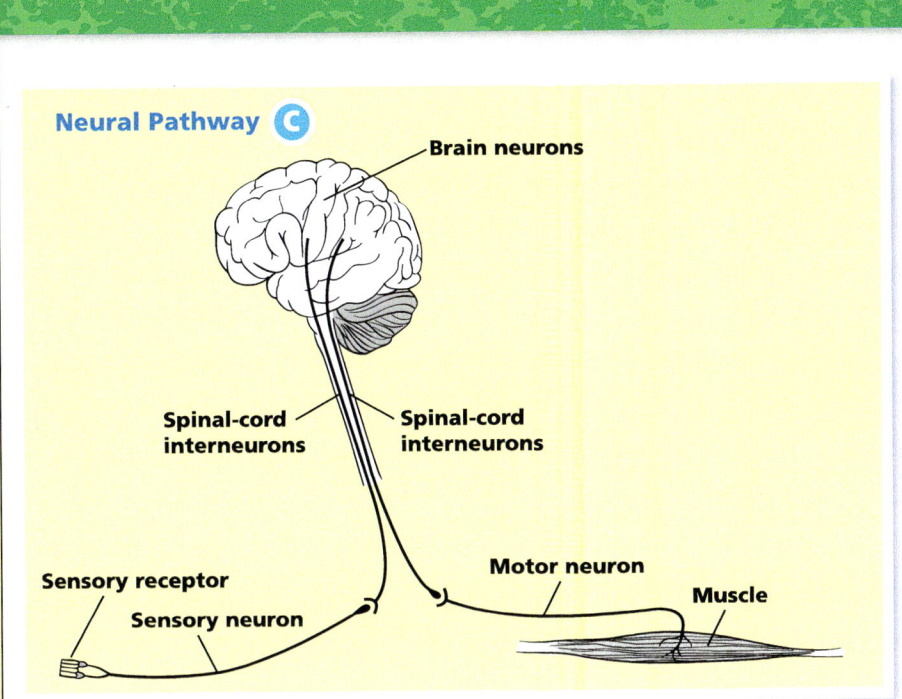

Neural Pathway

The brain receives and sends the message that leads from the sensation to the response.

Neural Pathways

The sensory receptor cells for all the senses connect to the ends of sensory neurons. A signal travels from the sensory receptor to a sensory neuron, which connects to **interneurons** (*inter-* means "between") in the spinal cord. Where do the interneurons carry the message? To the brain.

The brain processes the message and decides what to do. If you want to move your arm to brush off an ant, the brain sends out messages to the muscles in your arms. The message travels down interneurons in the spinal cord to **motor neurons**. Where do the interneurons carry the message? Away from the brain, to motor neurons in the arm, and finally to the muscles. Your arm moves.

Take Note

Use the information in these paragraphs to label the structures on your *Neural Pathways* notebook sheet.

Making sense of system models: Give students a few moments to review their *Neural Pathways* notebook sheet, then read aloud or call on a volunteer to read this section while the rest of the class follows the process on their notebook sheets adding in arrows and notes, and making sure they've labeled all the structures mentioned.

If needed, project a copy of the *Neural Pathways* notebook sheet and call on Reporters from each group to describe the functions as you record them.

CROSSCUTTING CONCEPTS

Systems and system models

ELA CONNECTION

This suggested strategy addresses the Common Core State Standards for ELA for literacy.

RST 4: Determine the meaning of symbols, key terms, and other domain-specific words and phrases.

Human Systems Interactions Course—FOSS Next Generation

INVESTIGATION 3 – The Nervous System

D

Use evidence to support explanation: Have students reread this section and compare their prior ideas about the inputs and outputs of the neural pathway system with the explanation given here. Encourage them to use evidence from the text to support their ideas. Students can also add any new information and/or connections they've made to their notes.

Encourage students to discuss the neural pathway in terms of a system. Ask,

➤ What are the parts of the system? Are there subsystems within this system?

➤ Explain what you think are the inputs and outputs of this system.

ELA CONNECTION

This suggested strategy addresses the Common Core State Standards for ELA for literacy.

RST 1: Cite evidence to support analysis of science texts.

WHST 9: Draw evidence from informational texts to support analysis, reflection, and research.

Nerve impulses travel through massive networks of neurons. These impulses control virtually everything in the body. The brain coordinates it all.

Communication

Sensory neurons, motor neurons, and interneurons are all like wires carrying an electric signal. Sensory neurons carry information *to* the brain, and motor neurons carry messages *from* the brain. Interneurons are between the other neurons and the brain. The motor neurons carry instructions to the muscles. Your arm muscles respond by contracting. Stretch receptors in the muscle then give the brain information on how much the muscles are stretched or contracted.

This kind of communication is happening constantly between your brain and the rest of your body. The flow of electric signals along neurons allows you to sense the environment, make decisions, move, breathe, and so on.

82

186   Full Option Science System

Part 2: Sending a Message

One cubic millimeter of brain tissue contains 1 billion neural connections.

Reaction Time

Sending messages takes time. The longer the pathway, the longer it takes for a stimulus to produce a response. The interval between stimulus and response is sometimes called **reaction time**. You may have noticed this delay if you have ever stubbed your toe. You can see it being stubbed and hear the sound before you feel the pain! The pathway from your eyes and ears to your brain is much shorter than the pathway from your toes to your brain. So the sensory neurons in your eyes and ears get their messages to the brain before the sensory neurons in your toes can.

Investigation 3: *The Nervous System* 83

E
Read aloud or call on a volunteer to reread this section. Have students pause to think about their own experiences and then share with a partner another example of "reaction time." Have a few students share and/or have them add their experiences to their notebooks.

End by having students reflect on what they learned from the article. Here are some suggested prompts for writing in their notebooks:
"I used to think ___, but now I know that ___.
It surprised me that ___.
I'm still wondering about ___."

ELA CONNECTION

These suggested strategies address the Common Core State Standards for ELA for literacy.

RST 9: Compare and contrast the information gained from experiments with that gained from reading a text on the same topic.

WHST 9: Draw evidence from informational texts to support analysis, reflection, and research.

Human Systems Interactions Course—FOSS Next Generation **187**

INVESTIGATION 3 – The Nervous System

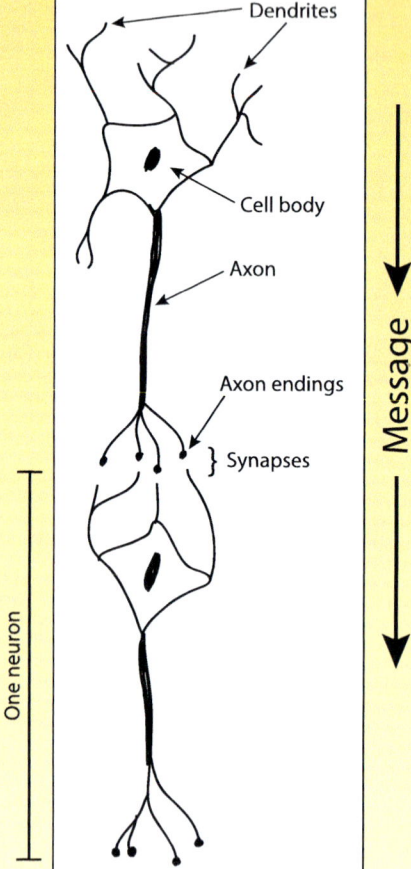

7. Preview the neural-message relay
Tell students that in the next session they will develop a model to explain how the neural pathway works. They will act out the functions of each structure in a neural-message relay. Each person will be a part of the pathway they have just outlined.

8. Introduce neurotransmission
Tell students that before they can perform the neural-message relay, they need to know more about how a message passes from one neuron to the next neuron. Say,

*Neurons do not actually touch each other. In order to keep a message moving through the nervous system, the message must cross a small gap between two neurons called a **synapse**. The message is carried across the synapse by chemicals called **neurotransmitters**.*

Make a rough sketch of two neurons on the board (see sidebar). Label one neuron with the cell parts and the synapse. Write "neurotransmitter" on the board. Explain that *neuro* means "nerve" or "nervous system" and *transmit* means "to send." Have students copy your diagram into their notebooks.

Project teacher master G, *Neural Transmission*, and point out the close-up of a synapse between an axon ending and a dendrite ending. Use the diagram as you continue.

A neuron's axon ends with a slightly swollen bulb. The dendrite of the next neuron also ends in a flatter-cupped bulb that is separated from the end of the axon by a small fluid-filled gap, the synapse.

When a message gets to the end of the axon, the axon bulb releases a chemical neurotransmitter (the red triangle). The neurotransmitters carry the message across the synapse to the dendrite of the next neuron.

The chemical neurotransmitter triggers the next neuron to fire and the electric message continues on its way.

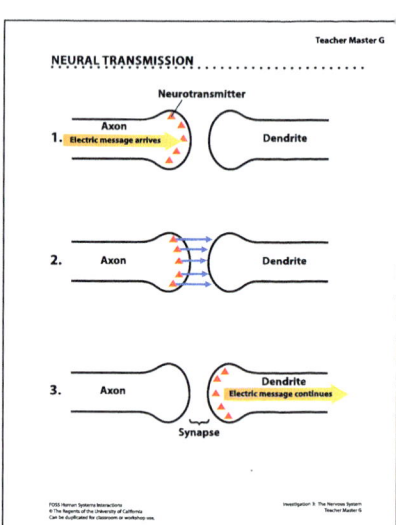

188   Full Option Science System

Part 2: Sending a Message

SESSION 2 *45–50 minutes*

9. Demonstrate the neural-message relay

Give students a few minutes to think about the activity in terms of a systems model. Say,

Together we are going to create a model to help us understand how a message passes from one neuron to the next. Imagine each of you is a neuron in this system. How might you represent the structures of a neuron?

Listen to students' ideas and suggest that to standardize the model, the right hand will be the dendrite, and the left hand will be the axon.

Tell students,

Imagine each of you is a neuron. Your right hand is a dendrite, and your left hand is an axon.

Hold up an aluminum-foil ball and a cup and say,

The function of this system is neurotransmission.

Give students time to discuss how they will represent the transfer of the neurotransmitter. Suggest that they can send the message to their neighbor by dropping the ball into a "dendrite receptor cup" of the neuron on your left side. Discuss other moves they need to consider in order for their model to be as accurate as possible.

➤ *When should you pass the neurotransmitter?* [Release the neurotransmitter only after you have received one in your dendrite-receptor cup in your right hand from your neighbor.]

Explain,

This is important: you cannot release your neurotransmitter until you have received one from your neighbor.

➤ *Which way will the electric message pass through you? Left hand to right hand or right hand to left hand?* [Messages move from dendrite to axon; the message moves from right hand to left.]

10. Distribute relay cards to the demonstration team

Tell students that you have a scenario they can use to test their system model. Ask two groups to make a big team of eight to serve as the demonstration team. Display the mosquito-stimulus card which you will hold onto, and then hand out the remaining eight relay cards to the team members and have them wear the cards as necklaces. Tell them that their first job is to organize themselves so that the steps of the neural pathway are in order from stimulus through to response. Give them a minute to work this out. Other students from the class can help by making suggestions.

SCIENCE AND ENGINEERING PRACTICES
Developing and using models

CROSSCUTTING CONCEPTS
Systems and system models

▶ **NOTE**
The eight demonstration team neurons and muscle should stand up and organize themselves in a circle in relation to the mosquito-stimulus card.

Human Systems Interactions Course—FOSS Next Generation

INVESTIGATION 3 — The Nervous System

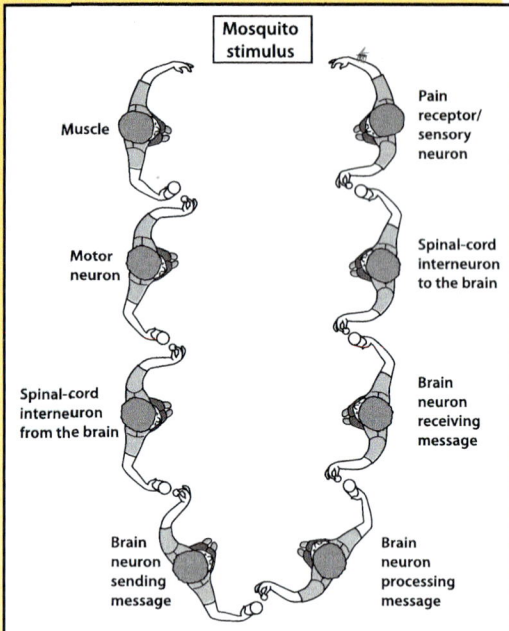

As a last resort, project teacher master K, *Neural-Message Relay Setup,* to provide direction.

The order should be (with the *mosquito stimulus* card between student 1 and student 8):

1. Pain receptor/sensory neuron
2. Spinal-cord interneuron to the brain
3. Brain neuron receiving message
4. Brain neuron processing message
5. Brain neuron sending message
6. Spinal-cord interneuron from the brain
7. Motor neuron
8. Muscle

11. **Pass out the cups and balls to demonstration team**
    Distribute the "neurotransmitters" and receptor cups to the demonstration team and describe the setup.
    - Each neuron will need a neurotransmitter (foil ball). The muscle doesn't need a foil ball.
    - Each neuron will need a receptor cup for a neurotransmitter except the pain receptor/sensory neuron.
    - Students will hold the neurotransmitter in the left hand (axon) and the receptor cup in their right hand (dendrite).
    - Students should be ready, with their left hand near the next person's receptor cup.
    - The stimulus to start the relay is a sting from the mosquito felt by the pain receptor/sensory neuron. You will call "Sting!"
    - No one drops the neurotransmitter into the receptor cup on the left until they have received the neurotransmitter from the person on the right.
    - The relay ends when the muscle swats the mosquito.

    Call "Sting!" and let the demonstration team begin the relay.

12. **Run the relay**
    Set up the other teams and distribute the materials. Give the teams a moment to arrange themselves in order. If there are extra students, give them tasks. One student could use a stopwatch to time a team. Other students could monitor the movement of the neurotransmitters.

CROSSCUTTING CONCEPTS
Systems and system models

TEACHING NOTE
You may want to run the demonstration in slow motion so all students can see and understand what is being done.

TEACHING NOTE
Students enjoy good-natured competition to see which team can relay the message the quickest. If you like, keep track of each class's best times to compare.

Part 2: Sending a Message

Call "Sting!" and watch the message travel from stimulus to response. Have the monitors watch for students who drop the neurotransmitters into the cups on their left before they have received one from their neighbor on the right.

Repeat the relay several times, inviting different students to be players and monitors.

▶ **NOTE**
The person serving as the "pain receptor" could also be the team timer and hold the stopwatch in their right hand.

13. Clean up
When the relays have been run, have students place the role cards, cups, and balls in the zip bags and return them to the materials station.

14. Debrief the relay
Give students a few minutes to discuss what they learned from the activity and how using a systems model helped them understand the concept of neurotransmission. They might also discuss the limitations of the model activity.

Hold a whole-class discussion using these questions as a guide.

➤ *Did you get faster as you ran more relays?* [Most likely yes.]

➤ *What affects the amount of time it takes to complete the relay?* [Length of pathway, number of connections, efficiency at passing and catching the neurotransmitter, repetition, and practice. New members of a team may not respond as quickly as experienced members.]

Students may have devised ways to make passing the neurotransmitter foil balls into the receptor cups faster, such as moving closer together and positioning the foil balls directly over the cups. These are all good adaptations to reduce response time.

➤ *What do you think affects the amount of time it takes for the brain and nerves to complete a real neural pathway from stimulus to response?* [Your focus, distractions, conditioning, health, age, the closer to your brain the sensory neuron is stimulated, and so on.]

Tell students that the body's neural network develops with age and that neurons that are frequently used in a pathway actually grow closer to each other as the pathway is reinforced.

15. Discuss neuron growth
Access the "Brain: Neuron Growth" online images. Click Play and project the images so students can observe the growth of neural networks over time (from newborn to 2 years old). Ask students to comment on anything they observed that was surprising.

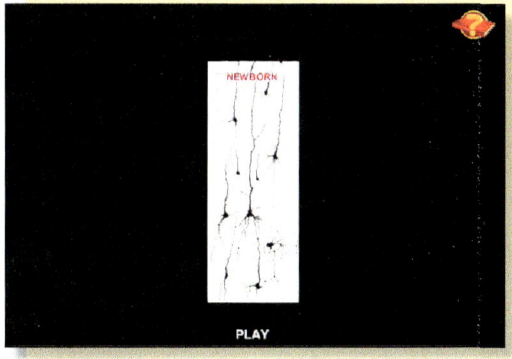

Human Systems Interactions Course—FOSS Next Generation

INVESTIGATION 3 – The Nervous System

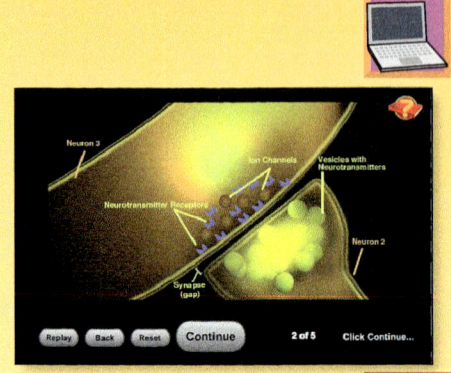

SCIENCE AND ENGINEERING PRACTICES

Analyzing and interpreting data

Engaging in argument from evidence

CROSSCUTTING CONCEPTS

Systems and system models

16. View online activity

Project the "Brain: Synapse Function" online activity. This activity simulates the passage of neurotransmitters over the synaptic gap.

Replay the simulation a few times and ask students to look for new information that is presented in the simulation that they haven't discussed already.

This is a good activity to have students review in pairs. They can take turns narrating the process and asking each other any questions to clarify the process.

17. Assess progress: response sheet

Use notebook sheet 9, *Response Sheet—Investigation 3* for assessment. Assign the response sheet as homework, or give students time to respond to the prompt in class. Distribute a copy of the response sheet to each student, but don't have them tape or glue it into their notebooks at this time.

Collect the sheets and use them to consider students' thinking about how messages travel to and from the brain.

What to Look For

- *Students draw a simple model of the nervous system (similar to the neural pathway on notebook sheet 8 or on page 61 of the* FOSS *Science Resources book) and show the areas that might be involved.*

- *Students describe three possible causes:*
  Problems with functioning of sensory receptors
  Problems with sensory neurons going to the spinal cord and/or to the brain.
  Problems with the interpretation of the stimulus in the cerebral cortex of the brain.

Plan to spend 15 minutes outside of class to review a sample of student responses, using *Embedded Assessment Notes* to record your observations. After your review, return the notebook sheets to students to be taped or glued into their science notebooks.

If a next-step strategy is needed, have groups compare their responses and discuss. Generate a class list of possible causes from the group discussions.

Give students time to draw a line of learning below their original responses and make any necessary changes. (Alternatively, they can glue or tape their response sheets in their notebooks on the left side of a two-page spread and rewrite their answers on the blank right side.)

Part 2: Sending a Message

SESSION 3 45–50 minutes

READING *in Science Resources*

18. Read "Neurotransmission: The Body's Amazing Network"

The article "Neurotransmission: The Body's Amazing Network" will help students clarify information about neural pathways. Use the guide on the following pages to conduct an in-class reading of the article.

At breakpoints indicated in the reading guide, stop the class for a minidiscussion. Students should have their notebooks set up to take notes and self-stick notes on hand.

19. Use a reading comprehension strategy

The main strategy to use for this reading is to have students integrate the information in the text and the diagrams with the information they gathered by doing the neural-message relay simulation in class.

SCIENCE AND ENGINEERING PRACTICES

Obtaining, evaluating, and communicating information

ELA CONNECTION

These address the Common Core State Standards for ELA for literacy.

RST 7: Integrate quantitative or technical information expressed in words in a text with a version of that information expressed visually.

TEACHING NOTE

For more information on scaffolding literacy in FOSS, see the Science-Centered Language Development in Middle School chapter in Teacher Resources.

INVESTIGATION 3 – The Nervous System

Ⓐ Analyze author's purpose: Read the title aloud and ask students to share with a partner why they think the author chose this title. Read aloud or ask for a volunteer to read the rest of the first page.

ELA CONNECTION

These address the Common Core State Standards for ELA for literacy.

RST 6: Analyze the author's purpose in providing an explanation, describing a procedure, or discussing an experiment in a text.

Ⓐ Neurotransmission: The Body's Amazing Network

Inside your body is a living communication system. It is more complex than the most advanced computer network.

Your brain and nerves send and receive billions of electric and chemical messages every minute. They connect everything, from the depths of the brain to the remote toes. Let's take a closer look.

From Neuron to Neuron

Remember that neurons are cells that carry the messages within the nervous system. The messages they carry are electric impulses.

194 Full Option Science System

Part 2: Sending a Message

Neurons pass their messages from one to another. But they do not touch each other. Instead, a tiny gap called a **synapse** separates the axon of one neuron and the dendrite of the next neuron.

Once an electric impulse reaches the end of an axon, it stimulates the release of chemicals into the gap. These **neurotransmitters** flow across the synapse. They fit into surface receptors of the receiving dendrite.

The neurotransmitters trigger the continuation of the electric message in the receiving dendrite. The electric impulse races down the axon. The process repeats, relaying the message on its way.

Take Note

Trace the pathway of one electric impulse from the axon of one neuron to the dendrite of the next. State what is happening at each step.

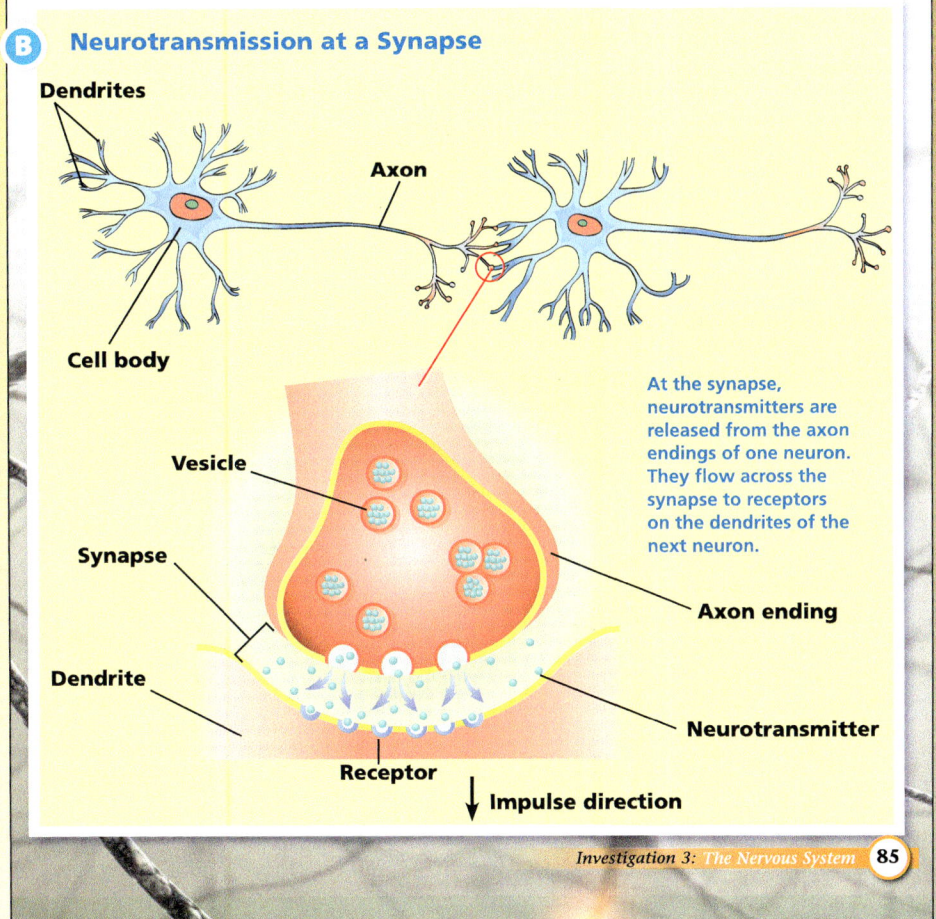

Neurotransmission at a Synapse

At the synapse, neurotransmitters are released from the axon endings of one neuron. They flow across the synapse to receptors on the dendrites of the next neuron.

Investigation 3: The Nervous System   85

Integrate technical information: Give students a few moments to look at the diagrams. Remind them of questions they can ask themselves to help them with comprehension, such as:

➤ *What is the purpose of this diagram? What does it show?*

➤ *What do the different shapes and arrows mean?*

➤ *What is the function of each structure represented?*

Cite evidence: Have students read this page independently to confirm their interpretations of the diagram. When they finish, have them turn and talk with a partner about anything new they learned from the text.

Have students take turns tracing the pathway of one electric impulse from the axon of one neuron to the dendrite of the next and state what is happening at each step.

ELA CONNECTION

This addresses the Common Core State Standards for ELA for literacy.

RST 1: Cite evidence to support analysis of science texts.

RST 7: Integrate quantitative or technical information expressed in words in a text with a version of that information expressed visually.

Human Systems Interactions Course—FOSS Next Generation   **195**

INVESTIGATION 3 – The Nervous System

Determine central idea of text: Have students read the rest of the article independently. Suggest they set up a notebook page to take two-column notes. One column can be for a summary of the text and the other column for what the text makes them think about (personal connection).

Summary	Personal Connection
What the article is about	What it makes me think about

Take note: Have students share with a partner an example from their own experience where a neural pathway gets stronger with practice and use.

ELA CONNECTION

This addresses the Common Core State Standards for ELA for literacy.

RST 2: Determine the central ideas or conclusions of a text; provide an accurate summary.

The Communication Network

The typical human brain has 100 billion neurons, and 1,000 trillion neural connections. This incredible number of pathways is responsible for everything a person learns, remembers, says, sees, and does. In fact, it is responsible for everything that makes a person human. A baby's brain is building those pathways even before it is born!

During the first few years of life, the brain continues to develop systems of interactions and communication among all its neurons. Every time a baby looks at something, information enters his or her brain and forms a pathway between neurons. With repetition, that pathway gets stronger and those neurons thrive. Neurons that are not used do not become connected in strong permanent pathways, and they eventually die.

An electroencephalogram (EEG) is a test that measures electrical activity in the brain. Even in infants, an EEG can detect abnormal brain waves and help doctors evaluate brain disorders.

86

196 Full Option Science System

Part 2: Sending a Message

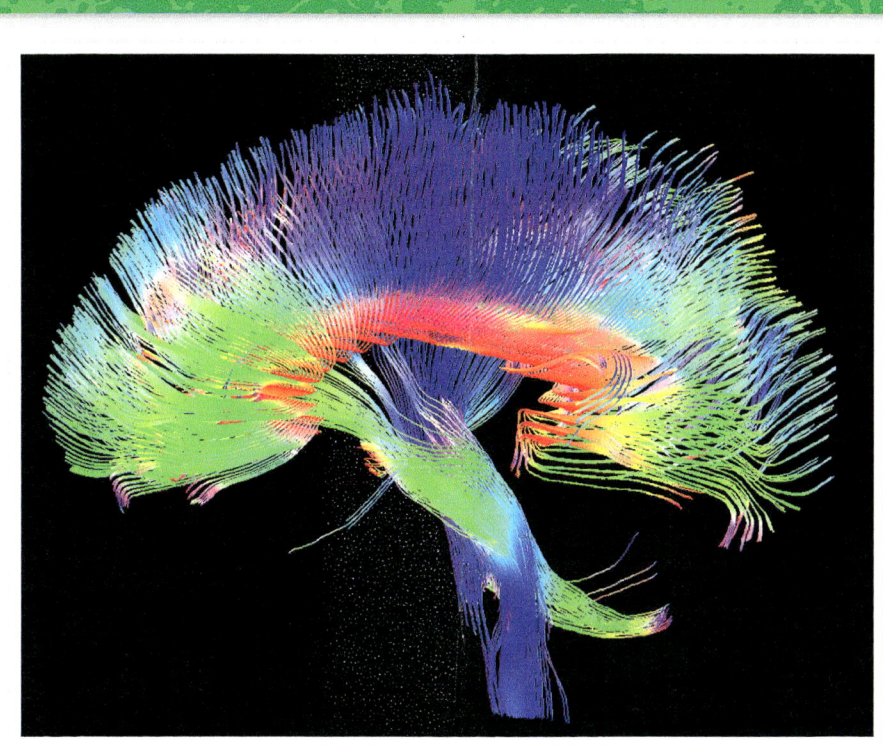

Every neuron in the brain is connected to as many as 15,000 other neurons, forming an incredibly complex network of neural pathways. The foundation for the network is laid in early childhood.

What does the formation of neural pathways have to do with memory formation?

Why do people say "use it or lose it"?

Many mysteries remain in our understanding of neural networks. For example, scientists are studying the birth and death of neurons. This work may lead to new treatments for brain diseases and disorders. Studies are showing that a teen's intellectual power is a match for an adult's. Your capacity to learn will never be greater than now. Build those neural pathways and take advantage of your brain power now!

Think Questions

1. The relay we did in class is a model for this process. Refer to the illustration on page 85. Compare the relay to the actual transmission of a message. What does each part of the relay represent?
2. Communication in the nervous system is called electrochemical. Why is this so?

Investigation 3: *The Nervous System* 87

Discuss the reading: Have students write down responses to these questions in their notebook. Have them share their responses with their group and cite evidence from the text to support their comparisons.

➤ *The relay we did in class is a model for this process. Compare the relay to the actual transmission of a message. What does each part of the relay represent?* [Scale in space and time is very different; foil ball represented neurotransmitter.]

➤ *Communication in the nervous system is called electrochemical. Why is this so?* [An electric impulse caused the chemicals at the synapse to release.]

ELA CONNECTION

These address the Common Core State Standards for ELA for literacy.

RST 9: Compare and contrast the information gained from experiments, simulations, video, or multimedia sources with that gained from reading a text on the same topic.

Human Systems Interactions Course—FOSS Next Generation

197

INVESTIGATION 3 — The Nervous System

cerebral cortex
nerve
neuron
neurotransmitter
synapse

20. Review vocabulary

Give students a few moments to review the vocabulary developed in this part. This is a good time for students to update their vocabulary indexes and tables of contents if they haven't already done so.

If students need more support, use a concept circle to help them understand the connections. Draw a circle on the board and/or have students make one in their notebooks. Divide the circle into quadrants and then write a vocabulary word in each one. Have students discuss in their groups the meaning and relationships between and among the words.

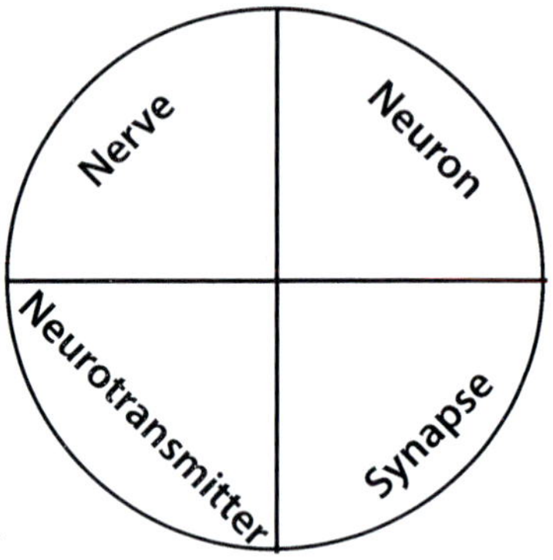

21. Answer the focus question

Tell students to write a response to the focus question. They should include a labeled drawing of two neurons with a synapse between them and describe the neural pathway from stimulus to response.

➤ *How do messages travel to and from the brain?*

Once students have completed their own responses, ask a few to share their thoughts with the class.

WRAP-UP/WARM-UP

22. Revisit the response sheet

Have students review their notebooks for the last few class sessions. Have them generate questions they have about the nervous system and discuss the questions with a partner.

EL NOTE

**Provide those students who need support with sentence frames such as "My diagram shows _____ .
First, _____ .
Then, _____ ."**

---

198 Full Option Science System

Part 3: Other Senses

MATERIALS for
Part 3: *Other Senses*

Provided equipment

For each student
- 1 *FOSS Science Resources: Human Systems Interactions*
- • "Sensory Receptors"
- • "Smell and Taste"
- • "Sight"

For each group
- 12 Paper clips, regular
- 2 Reaction-timer sheets (teacher master L)
- 1 Transparent tape

For the class
- 24 Vials with caps
- 24 Cotton balls
- 24 Removable labels
- 3 Cafeteria trays

Teacher-supplied items

For each group
- 1 Computer with Internet access
- 2 Scissors
- 2 Calculators
- 1 Mini-whiteboard and marking pen

For the class
- 8 Extracts or scented oils (see Getting Ready)
- 1 Crisp dollar bill

FOSSweb resources

For each student
- • Notebook sheet 10, *Smell*
- • Notebook sheet 11, *Reaction-Time Results*

For the class
- • Online activities: "Smell Menu," "Vision Menu," "Reaction Timer"

For the teacher
- • Teacher master L, *Reaction Timer*
- • Teacher master M, *Reaction-Time Questions*
- • *Embedded Assessment Notes*
- • Teaching slides, 3.3

FOSS Science Resources

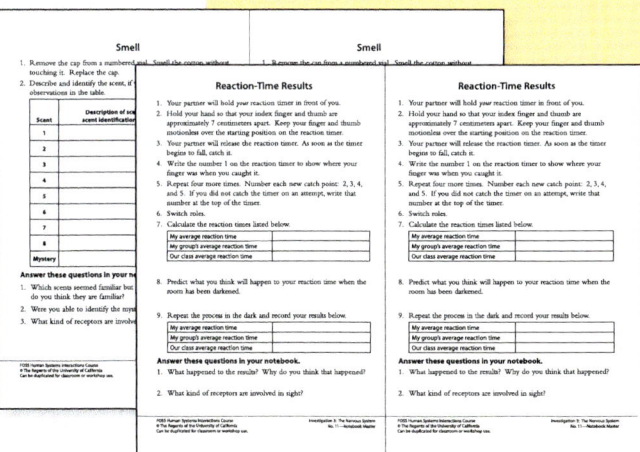

Nos. 10–11— Notebook Masters

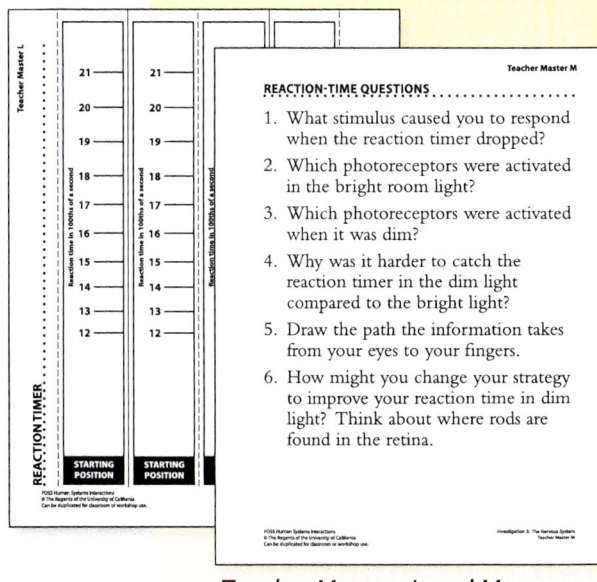

Teacher Masters L and M

Human Systems Interactions Course—FOSS Next Generation

199

INVESTIGATION 3 – The Nervous System

GETTING READY for
Part 3: Other Senses

Quick Start

Schedule	• Computers with Internet access, one for each group 3 active investigation sessions, including 2 readings
Preview	• Preview the FOSSweb Resources by Investigation for this part (such as printable masters, teaching slides, and multimedia) • Preview online activities: "Smell Menu," Step 9 "Vision Menu," Step 21 "Reaction Timer," Step 22 • Preview the reading: "Sensory Receptors," Steps 6, 19 • Preview/plan homework readings: "Smell and Taste," Step 10 "Sight," Step 26
Print or Copy	**For each student** • Notebook sheets 10, 11 **For each pair of students** • Teacher master L **For the teacher** • *Embedded Assessment Notes*
Prepare Material	• Prepare scent-identification vials **A** • Obtain a new $1 bill **B** • Construct sample reaction timer **C** • Prepare to darken the room **D**
Plan for Assessment	• Review Step 25, "What to Look For" in the notebook entry

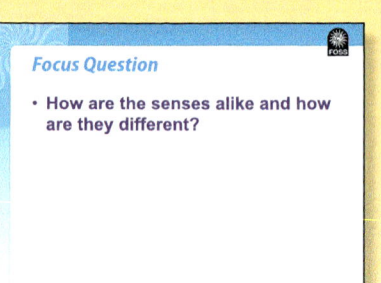

Teaching slides, 3.3

200 Full Option Science System

Part 3: Other Senses

Preparation Details

A ▶ Prepare scent-identification vials
You will need eight scents. Here are some suggestions.
- Almond
- Chocolate
- Cinnamon
- Citrus
- Cloves
- Coffee
- Eucalyptus
- Lavender
- Lemon
- Peppermint
- Pine
- Rose
- Sage
- Smoke
- Vanilla

Substitutions are fine. The local grocery store is a good source for many of the scents, though you may have to be creative to obtain herbal/spice scents such as cinnamon, cloves, and eucalyptus. Shops that sell body lotions are possible sources for the nonfood extracts.

You will need two sets of eight scent-identification vials and one set of eight mystery vials.

Label two sets of vials and caps with numbers 1–8, using the labels. Cut the labels in half, and use one half for a vial, and one half for its cap. Place one cotton ball in the bottom of each vial. Sprinkle or pour a very small amount of scent into each vial and snap the cap on. Use a copy of notebook sheet 10, *Smell*, to record which scent is in each vial. Place each set of scent-identification vials on a tray or in a box to keep them separate.

You will also need eight mystery vials. Prepare these vials in the same way you did the others, except label the vials and caps "M." Choose one of the eight scents to be the mystery scent and put it in each of the mystery vials. Put them aside until after the students have smelled the numbered vials.

B ▶ Obtain a new $1 bill
Get a $1 bill that is fairly new. You will use this in Step 11 to introduce reaction time.

C ▶ Construct sample reaction timer
Construct a reaction timer for yourself, so you understand how to use and record on it. See Step 13 of Guiding the Investigation for instructions. Each student will make his or her own reaction timer.

D ▶ Prepare to darken the room
In Step 16, you need to get the room as dark as possible. Ambient light through the windows might prevent it from being dark enough. You want students to have a hard time seeing their outstretched hands. Consider taping dark butcher paper over the windows to darken the room sufficiently.

▶ **SAFETY NOTE**
Be sure to find out if any students have allergies or sensitivities to extracts or scented oils.

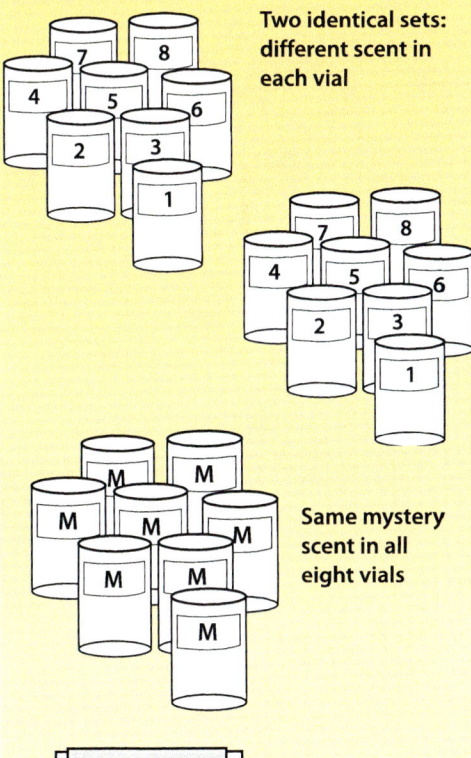

Two identical sets: different scent in each vial

Same mystery scent in all eight vials

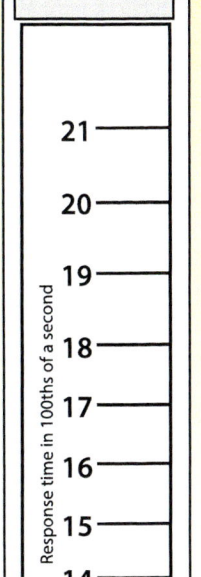

Human Systems Interactions Course—FOSS Next Generation

INVESTIGATION 3 – The Nervous System

FOCUS QUESTION

How are the senses alike and how are they different?

GUIDING *the Investigation*
Part 3: *Other Senses*

SESSION 1

Students will . . .
- Investigate their sense of smell with mystery scents (Steps 1–5)
- Read about chemoreceptors in "Sensory Receptors" (Steps 6, 7)
- Compare sensory receptors (Step 8)
- Gather information from online activity about how smell travels to the brain (Step 9)
- Read "Smell and Taste" as homework (Step 10)

SESSION 2

Students will . . .
- Discuss sight and a reaction-time demonstration (Steps 11, 12)
- Test and compare reaction time (Steps 13–15)
- Test and compare reaction time in a darkened room (Steps 16, 17)
- Discuss photoreceptors (Step 18)

SESSION 3

Students will . . .
- Read about and discuss photoreceptors in "Sensory Receptors" (Steps 19, 20)
- Gather information from online activity about vision sensory pathway to apply to reaction time (Steps 21, 22)
- Review vocabulary and answer focus question (Steps 23–25)
- Read "Sight" as homework (Step 26)
- Discuss research on sight (Step 27)

SESSION 1   45–50 minutes

1. **Introduce *smell* and *vision***
   Remind students that they have studied how the nervous system functions in the sense of touch. They will now turn their attention to two other senses, **smell** and **vision** (or sight), to see if they operate in a manner similar to the sense of touch.

EL NOTE

If necessary, refer to the "Senses" word wall from Part 1 to point out "smell" and "vision."

Part 3: Other Senses

2. **Focus question: How are the senses alike and how are they different?**

    Pose the focus question, write or project it on the board, and have students write it in their notebooks.

    ➤ *How are the senses alike and how are they different?*

    Students do not need to answer the question at this time. They should leave the rest of the page blank to return to the question at the end of this part.

3. **Discuss smell**

    Ask students to share with a partner their responses to the following questions, before sharing with the class. Expect students to have strong reactions!

    ➤ *What is your favorite smell?*

    ➤ *What is your least favorite smell?*

    Tell students that the sense of smell is quite important.

    Natural gas is an odorless, colorless gas that is used throughout most of the country to cook and heat homes. It is also dangerously explosive.

    ➤ *How do you think that the gas companies have made this gas safer?* [A strong-smelling chemical is added to the gas so that it can be identified by its unique smell.]

    ▶ **SAFETY NOTE**
    Be sure to find out if any students have allergies or sensitivities to extracts or scented oils.

4. **Conduct the smell activity**

    Distribute notebook sheet 10, *Smell*, to each student. Describe the activity and how you want students to pass the numbered scent vials. Divide the class in half (each half will have about 16 students). Each half of the class will receive a set of eight scent vials that they will share. Students should smell the vials, record their observations, and enter a guess about identification. Continue until everyone has smelled all eight vials.

    TEACHING NOTE
    Expect high levels of identification of some scents and surprise on others.

5. **Debrief smell activity and introduce mystery scent**

    After students have smelled scents 1–8, poll the class regarding the identification of each smell. Confirm or identify the scents, listen to their consternation, and have them answer question 1 on the notebook sheet.

    Tell students that you have a mystery scent that is the same as one of the scents they have already smelled. Distribute four mystery scent vials to each half of the class and ask students to identify it. They should answer question 2. Give them time to share their determinations.

    TEACHING NOTE
    Note that sensory adaptation and confusion occur easily in the olfactory system. Students will vary in their ability to identify the mystery scent.

Human Systems Interactions Course—FOSS Next Generation

INVESTIGATION 3 – The Nervous System

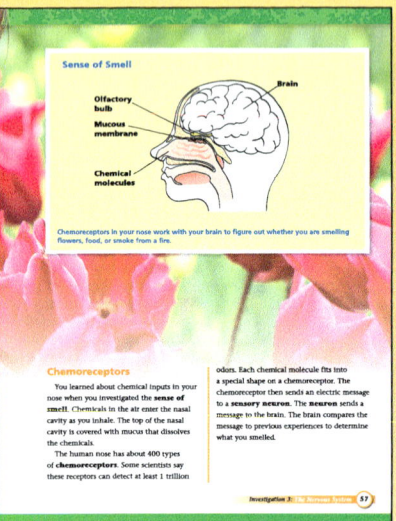

READING *in Science Resources*

6. Read "Sensory Receptors—Chemoreceptors"

Start by having students think about how smell is sensed. Give them a few moments to share ideas with their group and then hold a class discussion using these questions as a guide.

➤ *Does your nose have the same kind of sensory receptors as your skin?* [Probably not. Skin has mechanoreceptors, but smell is different because it's not a touch.]

➤ *What do you think stimulates the sensory receptors in your nose?* [A smell.]

➤ *What is a smell and how does it get to your nose?* [Smell is a chemical particle and it travels through the air.]

Say,

In order for you to smell something, molecules from that thing have to make it to your nose. The molecules travel through the air, up your nose to special receptors called **chemoreceptors**. *Let's see how this works.*

7. Use a reading comprehension strategy

Ask students to turn to the article "Sensory Receptors" in *FOSS Science Resources*. Remind them that they read about mechanoreceptors a short time ago. Refer to the concept grid started in Part 1 and tell students that now they will find out more about chemoreceptors. They should continue to gather and organize information using the concept grid.

Use the guide on the following pages to conduct an in-class reading of the "Chemoreceptors" section of the article.

ELA CONNECTION

This suggested strategy addresses the Common Core State Standards for ELA for literacy.

SL 1: Engage in collaborative discussions.

SCIENCE AND ENGINEERING PRACTICES

Obtaining, evaluating, and communicating information

TEACHING NOTE

For more information on scaffolding literacy in FOSS, see the Science-Centered Language Development in Middle School chapter in Teacher Resources.

---

204   Full Option Science System

Part 3: Other Senses

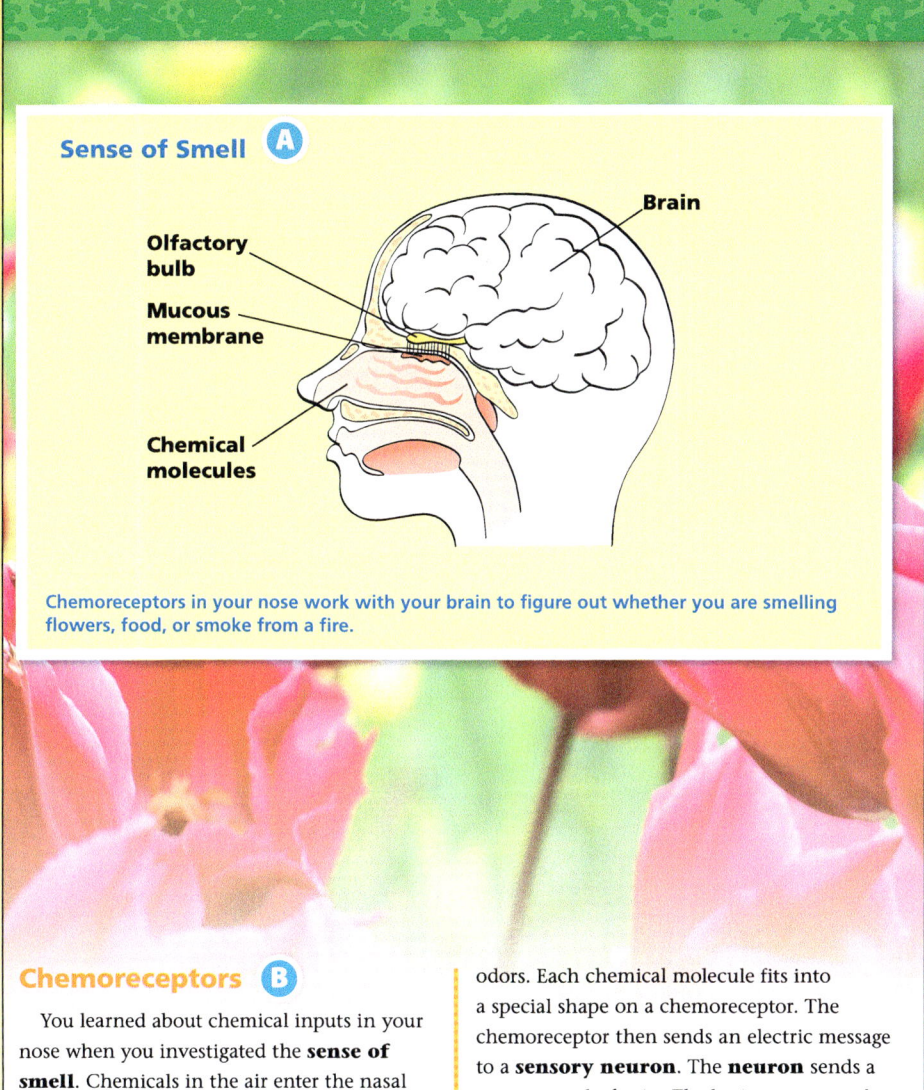

Sense of Smell

Chemoreceptors in your nose work with your brain to figure out whether you are smelling flowers, food, or smoke from a fire.

Chemoreceptors

You learned about chemical inputs in your nose when you investigated the **sense of smell**. Chemicals in the air enter the nasal cavity as you inhale. The top of the nasal cavity is covered with mucus that dissolves the chemicals.

The human nose has about 400 types of **chemoreceptors**. Some scientists say these receptors can detect at least 1 trillion odors. Each chemical molecule fits into a special shape on a chemoreceptor. The chemoreceptor then sends an electric message to a **sensory neuron**. The **neuron** sends a message to the brain. The brain compares the message to previous experiences to determine what you smelled.

Investigation 3: The Nervous System — 57

A

Integrate technical information: Have students look at and interpret the diagram. Ask them to share with a partner what they recognize in the diagram and what is new to them.

B

Determine central ideas: Have students read pages 57 and 58 using a partner/share strategy. Partner A reads page 57 while Partner B listens and follows along. Then Partner B summarizes. Partner A makes sure the summary includes the important points. Partners switch roles for page 58.

ELA CONNECTION

These suggested strategies address the Common Core State Standards for ELA for literacy.

RST 2: Determine the central ideas or conclusions of a text; provide an accurate summary.

RST 7: Integrate quantitative or technical information expressed in words in a text with a version of that information expressed visually.

Human Systems Interactions Course—FOSS Next Generation 205

INVESTIGATION 3 – The Nervous System

Integrate technical information: Encourage students to use the diagram of the taste bud to help them summarize this section of the text.

D

Cite evidence: Give students a few moments to discuss the Think Question. Encourage students to cite evidence from the text to support their ideas.

➤ *Why can't you taste food very well when you have a cold?* [Your brain combines information from taste buds and nose to produce taste. If your nose has too much mucous from a cold, your smell receptors can't function.]

Call on a few volunteers to share their thoughts or those of their partner.

ELA CONNECTION

These suggested strategies address the Common Core State Standards for ELA for literacy.

RST 1: Cite evidence to support analysis of science texts.

RST 7: Integrate quantitative or technical information expressed in words in a text with a version of that information expressed visually.

The **sense of taste** also uses chemoreceptors. Look at a partner (or in a mirror). Stick out your tongue. Notice all the little bumps on your tongue? There are taste buds along the edges of the bumps—about 10,000 taste buds total.

Some chemicals in food dissolve in saliva. They touch the tongue's taste buds. The sides of the taste buds have different chemoreceptors that respond to one basic taste. Those tastes are salty, sweet, sour, bitter, and umami. Umami (meaty) was the most recently discovered taste. Scientists think that there might be a sixth taste—for fats.

The taste-bud neurons send messages to the brain. The brain compares them to previous experiences to determine what was tasted. Your brain combines a food's taste, smell, and texture into the sensation of flavor. You can see how smell affects taste by plugging your nose the next time you eat something.

Think Question

Why can't you taste food very well when you have a cold?

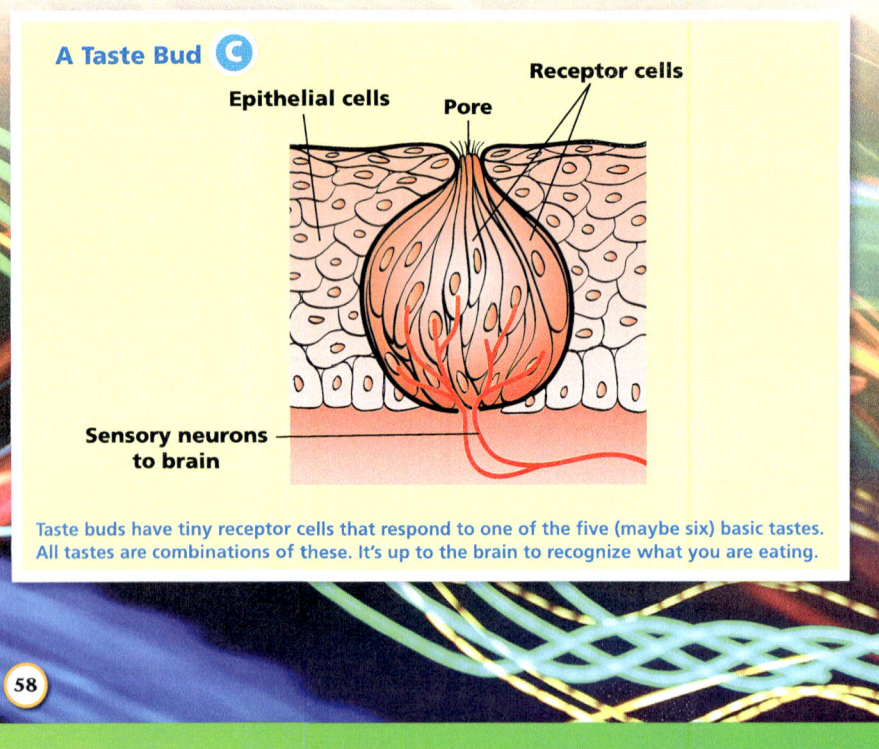

Taste buds have tiny receptor cells that respond to one of the five (maybe six) basic tastes. All tastes are combinations of these. It's up to the brain to recognize what you are eating.

58

206 Full Option Science System

Part 3: Other Senses

8. **Compare sensory receptors**

    Record new information on the Sensory Receptors concept grid in the chemoreceptors column. (This can be done with the whole class or students can work in groups or pairs.)

    Review the mechanoreceptors column and then have students do a comparison. Ask,

    ➤ *Compare and contrast chemoreceptors and mechanoreceptors.* [Both receive information from the environment and send it to the brain. Chemoreceptors respond to chemicals instead of pressure or motion. Chemoreceptors are found in the nose and mouth; mechanoreceptors are found in the skin and ears.]

    Students should answer the final question on the *Smell* notebook sheet.

9. **View online activity: "Smell Menu"**

    Project the online activity "Smell Menu." Use "Sensory Pathway" to demonstrate how smell stimuli travel to the brain. If students have access to computers, this is a good time to use them.

    Tell students to open "Receptors" and step through it. Instruct students to engage each activity several times on their own. They should click the Information button at the upper right and answer the questions in their notebooks. Students can explore the "Common Cold" activity if they wish.

    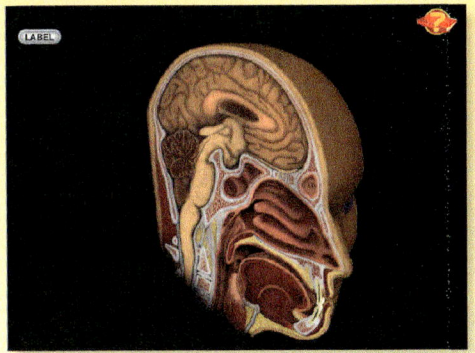

10. **Extend the investigation with homework**

    The article "Smell and Taste" is a good homework reading. Ask students to read the article and write down the information that they find most interesting.

    Suggest students set up their notebooks to take notes using a 3-2-1 strategy. As they read, they can write down 3 things they found most interesting; 2 important ideas; and 1 question (see example).

3	things that are most interesting
2	important ideas
1	question I have

    Students should also write down their response to the Take Note prompt in their notebooks.

Human Systems Interactions Course—FOSS Next Generation

INVESTIGATION 3 – The Nervous System

SESSION 2 45–50 minutes

11. Discuss sight

Tell students that they will now shift their attention to another sense, the sense of vision, also known as sight. Call on a volunteer to try an experiment with a dollar bill. Tell the student,

We all know money is hard to hold onto. Your challenge is to catch the dollar. I will hold the dollar, and you will position your fingers with index finger and thumb in the ready position about 6–7 cm apart, near the bottom of the bill. You catch the bill when you see it start to fall. No touching the bill while in the ready position, and a false start is a disqualification.

Try the drop with a student or two (though all will want to try!). They may able to respond quickly enough to catch the dollar, but that is rare. Encourage them to engage in this activity at home.

12. Discuss results

Have students discuss why the volunteer didn't catch the dropped dollar. Then call on a few volunteers to share what they or their partner thought.

If it didn't come up in the discussion, give students a moment to recall what they read in the "Brain Messages" article. Confirm that, after the eyes perceive the onset of motion, time is required to notify the brain, for the brain to process the information, and for the brain to send a message to the muscles to react.

Tell students,

The time it takes for a stimulus to produce a response is called **reaction time**. *From the moment you see something until you act on that visual signal requires time, but how much time?*

Tell students that they are going to investigate reaction time.

13. Introduce the reaction timer

Explain that you don't have a way to measure the brain's electric output. Ask students to think about and discuss in their groups a method for measuring how long it takes the body to receive and respond to a visual stimulus. Call on Recorders to share their group's idea.

Hold up a reaction timer. Tell students that this is called a reaction timer and that they will use it much like you used the dollar bill. Distribute two copies of teacher master L, *Reaction Timer*, to each group.

TEACHING NOTE

We use the terms "sense of sight" and "vision" interchangeably. After introducing the terms, do so yourself. Add "sight" next to the word "vision" on the senses word wall.

EL NOTE

Write "stimulus" and "response" on the board or word wall. Between the two words, draw an arrow and write the word "reaction time."

---

208 Full Option Science System

Part 3: Other Senses

Give the following instructions.

a. Cut on the dashed lines. Each sheet makes two timers.

b. Fold the reaction timer in half on the solid line and tape the top edge together.

c. Attach three paper clips along the bottom edge.

Tell students they will work in pairs and test each other. Distribute notebook sheet 11, *Reaction-Time Results*, and go through the procedure together. You might want to demonstrate with a student. Point out that the timer is calibrated in 100ths of a second.

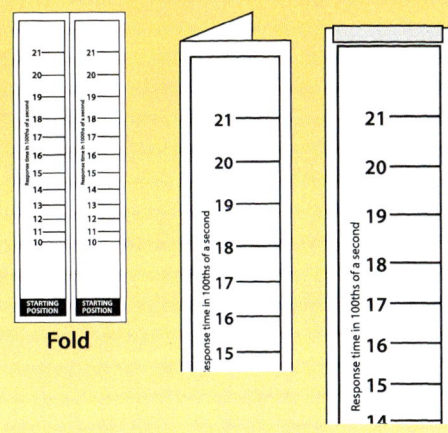

14. Test reaction time

Have students get their materials, construct their reaction timers, and begin testing. As students do their testing, cruise the groups and make sure that students are using the correct testing procedure and recording their results on the reaction timer as described in Steps 4-5 on notebook sheet 11.

15. Compare results

When students have done their testing and have calculated their individual and group averages, ask them to report their group averages to the class. Use the data to calculate a class average for reaction time. You might want to post a chart for classes to compare their reaction-time averages. Ask,

➤ What was the average length of time it took for a reaction? [0.15–0.21 second.]

➤ Why do you think it takes that long to react to a visual stimulus? [The message has to travel from the eyes to the brain, to be processed in the brain, and to travel from the brain to the fingers.]

16. Darken the room

Ask,

➤ What will happen to the average reaction time if the room is darkened and it is hard to see the timer? [It might take longer to react.]

Tell students to write down their prediction on their notebook sheet (item 8) and then to try it out. They should follow the same process as before and record the results of five trials on their reaction timer. They need to figure out a way to distinguish between the first set of numbers tested in the light and the second set of numbers they will test in darkness. Let students decide on a way, either on their own or as a class. Turn off the lights to make the room sufficiently dark for students to have a difficult time seeing the initial drop. Let the testing continue.

> **TEACHING NOTE**
>
> If students do not catch the timer on an attempt, they should write the number of that attempt at the top of the timer and not include it in the average.
>
> If students catch the timer before the 10, they can simply mark the spot with an X and you can help them infer the appropriate time to include for the average.

> **TEACHING NOTE**
>
> Two possible ways of marking the new trials are to write the results on the opposite side of the reaction timer or to signify the new results with the letter D. If students can't come to a reasonable solution on their own, suggest one of these.

> **TEACHING NOTE**
>
> If you can't sufficiently darken the room or you question the results, ask students to consider what might happen to the results in a completely dark room, one with no light whatsoever.

Human Systems Interactions Course—FOSS Next Generation

INVESTIGATION 3 – The Nervous System

SCIENCE AND ENGINEERING PRACTICES

Analyzing and interpreting data

Constructing explanations

E L   N O T E

Refer to the class Sensory Receptors concept grid to review and then point out or add "photoreceptors."

---

17. Discuss results

Ask students what they found out. For the most part, they should respond that it was much more difficult to see the reaction timer and therefore they could not catch it as easily. You can determine if it is worth calculating a class average.

Ask students to respond to the questions on the notebook sheet and then ask them to share their thoughts. Students might respond that the brain didn't get the message as quickly or that the receptors in their eyes don't work as well when it's dark.

18. Introduce *photoreceptor*

Turn the conversation to the sensory receptors. Ask,

➤ *What kinds of sensory receptors have we investigated so far?* [Mechanoreceptors such as pressure and pain receptors, and chemoreceptors in the nose.]

➤ *Do the receptors in your eyes respond to the same stimuli as mechanoreceptors or chemoreceptors?* [No.]

➤ *What stimulates the sensory receptors in our eyes?* [Light.]

➤ *What kind of receptors are they?* [Students may not know.]

Tell students that the sensory receptors in our eyes are sensitive to light, rather than pressure or chemicals. Ask,

➤ *What do you call a person who uses a camera to take pictures?* [A photographer.]

Tell students that the root *photo* means light. The receptors in our eyes are sensitive to light. They are called **photoreceptors**.

Part 3: Other Senses

SESSION 3 45–50 minutes

READING *in Science Resources*

19. Read "Sensory Receptors"
Have students turn to the article, "Sensory Receptors." Remind them that they read about mechanoreceptors and chemoreceptors and now they will read about photoreceptors.

20. Use a reading comprehension strategy
Refer to the concept grid and tell students that they should continue to gather and organize information using the concept grid.

Use the guide on the following pages to lead an in-class reading of the section titled "Photoreceptors." Read this section together.

Begin by having students analyze the diagram. They should ask themselves: How does this diagram compare to other sensory receptor diagrams? What do I already know about the eye and what questions do I have?

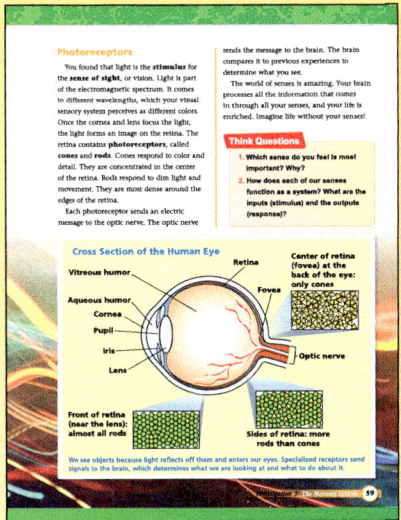

ELA CONNECTION

This suggested strategy addresses the Common Core State Standards for ELA for literacy.

RST 7: Integrate quantitative or technical information expressed in words in a text with a version of that information expressed visually.

TEACHING NOTE

For more information on scaffolding literacy in FOSS, see the Science-Centered Language Development in Middle School chapter in Teacher Resources.

Human Systems Interactions Course—FOSS Next Generation **211**

INVESTIGATION 3 – The Nervous System

Compare and contrast: Tell students to read this section independently to find out about the two types of photoreceptors.

Discuss what students found out about cones and rods. [Cones are sensitive to bright light, color, and detail. Rods function in dim light and can see shape but not color.]

Tell students to write down the difference between rods and cones in their notebooks.

Discuss the reading: Have groups discuss the think question at the end of the article. They should describe how each sense functions as a system. Students should focus on the inputs (stimulus) and outputs (response).

ELA CONNECTION

This suggested strategy addresses the Common Core State Standards for ELA for literacy.

RST 9: Compare and contrast the information gained from experiments, simulations, video, or multimedia sources with that gained from reading a text on the same topic.

Photoreceptors

You found that light is the **stimulus** for the **sense of sight**, or vision. Light is part of the electromagnetic spectrum. It comes in different wavelengths, which your visual sensory system perceives as different colors. Once the cornea and lens focus the light, the light forms an image on the retina. The retina contains **photoreceptors**, called **cones** and **rods**. Cones respond to color and detail. They are concentrated in the center of the retina. Rods respond to dim light and movement. They are most dense around the edges of the retina.

Each photoreceptor sends an electric message to the optic nerve. The optic nerve sends the message to the brain. The brain compares it to previous experiences to determine what you see.

The world of senses is amazing. Your brain processes all the information that comes in through all your senses, and your life is enriched. Imagine life without your senses!

Think Questions

1. Which sense do you feel is most important? Why?
2. How does each of our senses function as a system? What are the inputs (stimulus) and the outputs (response)?

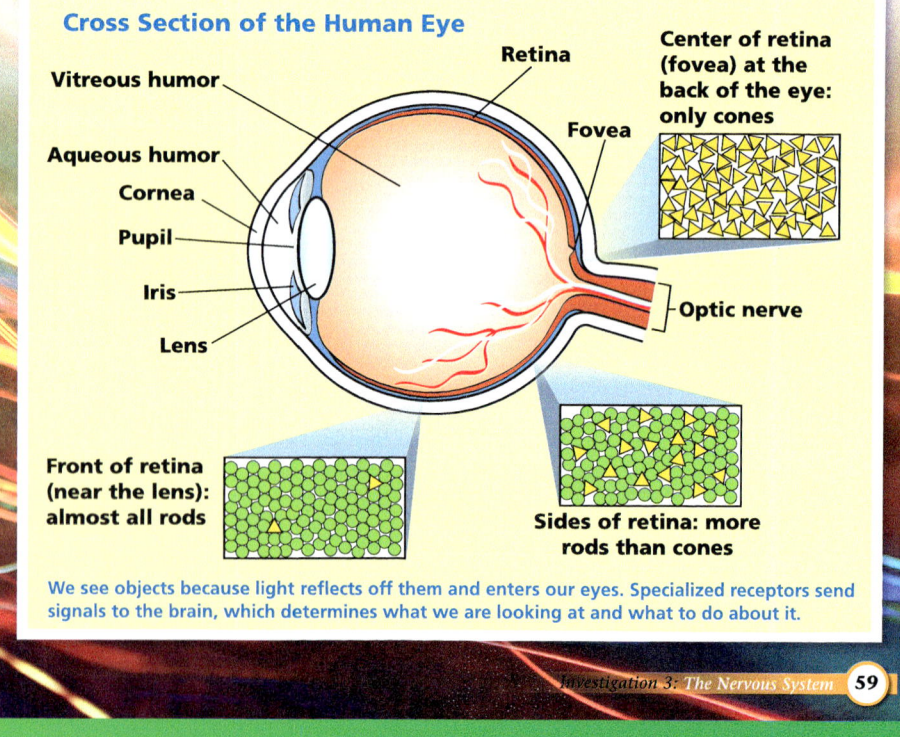

Cross Section of the Human Eye

We see objects because light reflects off them and enters our eyes. Specialized receptors send signals to the brain, which determines what we are looking at and what to do about it.

Part 3: Other Senses

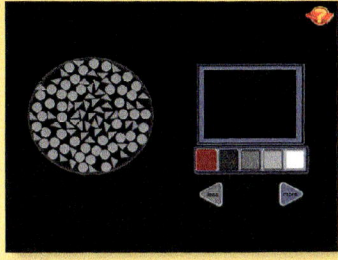

21. **View online activities: "Vision Menu"**

    Access the online activity "Vision Menu." Make sure the sound is turned on. Project the activity "Sensory Pathway" for students and follow the path of the sensory information from the eyes to the back of the brain where vision information is processed.

    Project the online activity "Receptor Density." Step through the amount of light, asking students to determine which shape of receptor represents rods and which represents cones. They should be able to do this based on the amount of light present.

22. **Return to reaction-timer activity**

    Tell groups that they will work together to think about and discuss what they've learned so far about reaction time and photoreceptors. Assign roles such as Facilitator, Questioner, Recorder, and Reporter. Project teacher master M, *Reaction-Time Questions*, and go through the questions one by one. The Questioner will pose the questions one by one to the group. The Facilitator makes sure everyone's ideas are heard. The Recorder writes down the group's responses. If you have mini-whiteboards and whiteboard marking pens, distribute one to each Recorder. When groups are finished, call on the Reporters to share their groups' responses.

    SCIENCE AND ENGINEERING PRACTICES
    Obtaining, evaluating, and communicating information

    ▶ *What stimulus caused you to respond when the reaction timer dropped?* [Light bouncing off the paper; seeing movement.]

    ▶ *Which photoreceptors were activated in the bright room light?* [Cones.]

    ▶ *Which photoreceptors were activated when it was dim?* [Rods.]

    ▶ *Why might be it harder to catch the reaction timer in the dim light compared to the bright light?* [It was hard to see the strip in the dark room.]

    ▶ *Draw the path the information travels from your eyes to your fingers.* [Photoreceptors to optic nerve to back of the brain to the spinal cord to motor neurons, to the muscles in the arm that control the hand.]

    TEACHING NOTE
    Note that the sensory and processing part of this information was seen in "Vision Menu: Sensory Pathway" activity.

    ▶ *How might you change your strategy to improve your reaction time in dim light? Think about where rods are found on the retina.* [Rods are concentrated on the sides of the retina, resulting in peripheral vision (what you detect from the side of your eye). Peripheral vision is good at detecting motion in dim light. Instead of focusing your eyes straight on the paper strip, you could turn your head and use your peripheral vision to see the reaction time. Another approach is to give your eyes time to adjust to the dim light so that more rods become active.]

Human Systems Interactions Course—FOSS Next Generation

INVESTIGATION 3 – The Nervous System

chemoreceptor
photoreceptor
reaction time
smell
vision

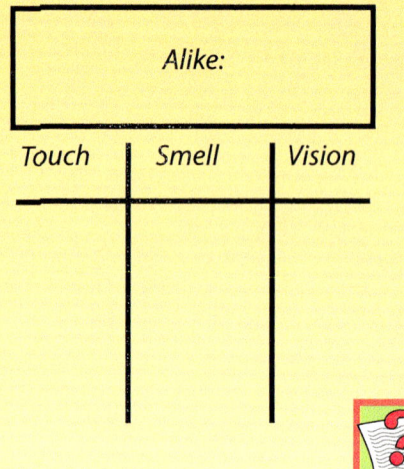

SCIENCE AND ENGINEERING PRACTICES
Constructing explanations

CROSSCUTTING CONCEPTS
Systems and system models

23. Record vocabulary
Give students a few moments to review the vocabulary developed in this investigation. They should highlight the vocabulary words, add to their labeled diagrams, and update their vocabulary indexes if they haven't already done so.

24. Answer the focus question
Ask students to summarize their understanding of the three senses they have investigated: touch, smell, and sight. If you would like, they can include hearing and taste, as well.

➤ *How are the senses alike and how are they different?*

Have students create a diagram to organize their understanding. This could be a box-and-T chart, or a Venn diagram, or another presentation. Tell students to use their notebooks, including the concept grid comparing sensory receptors, and *FOSS Science Resources* for assistance. This activity could take some time. When they are done, give students time to share their responses. Using a line of learning or another color, they can add to their original responses.

25. Assess progress: notebook entry
Review the diagram that students created to answer the focus question to check their understanding about how senses are alike and different. Tell students to include specific examples in their response.

What to Look For

- *Similarities include examples such as all the senses have sensory receptors that send messages via neurons to the brain; all the senses allow a human to gather information from the environment. Each sense functions as a system with inputs and outputs.*

- *Different sensory receptors respond to different stimuli (mechanoreceptors respond to pressure or movement, chemoreceptors respond to chemicals, photoreceptors respond to light).*

- *Includes examples, (e.g., mechanoreceptors are located in the skin; photoreceptors are located in eye, chemoreceptors are located in the nose).*

Record your observations on a copy of *Embedded Assessment Notes*. Plan next-step strategies as needed.

Part 3: Other Senses

26. Extend the investigation with homework

Students can read the article "Sight" for homework. As they read, tell them to write down questions on aspects of the reading they find most interesting. Have them choose one of those questions to research further using other sources, such as the FOSSweb multimedia resources on vision. Their task is to bring back one "fun fact" to share and discuss with their group.

You might consider having students read this before they answer the focus question, but it is not entirely necessary.

You can also have students try out the "Reaction Timer" online activity at home. They should go through both phases of the online activity, and think about the implication of the results. You might ask them to print out the last page of the activity and write a response to what they think the results mean on the same paper.

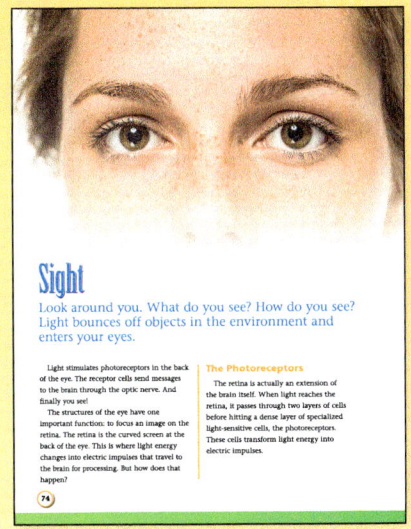

WARM-UP

27. Discuss research on sight

The next day, spend a few minutes discussing the "fun fact" students discovered in their reading and research on sight. Have students share those in their groups and then have a few groups share a fact.

ELA CONNECTION

This suggested strategy addresses the Common Core State Standards for ELA for literacy.

WHST 7: Conduct short research projects to answer a question.

Human Systems Interactions Course—FOSS Next Generation   215

INVESTIGATION 3 – The Nervous System

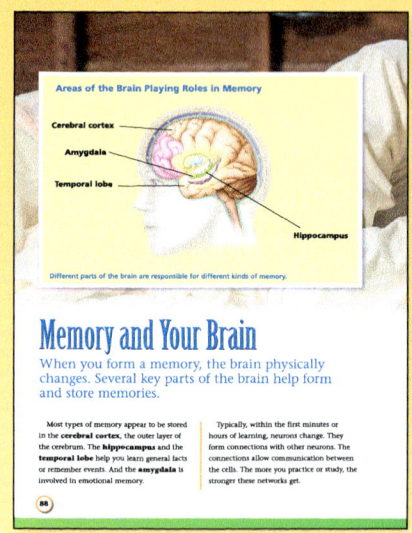

FOSS Science Resources

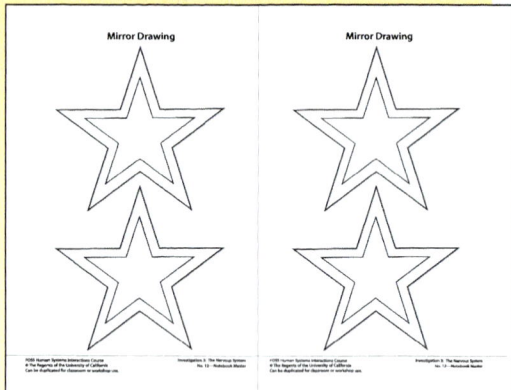

No. 12—Notebook Master

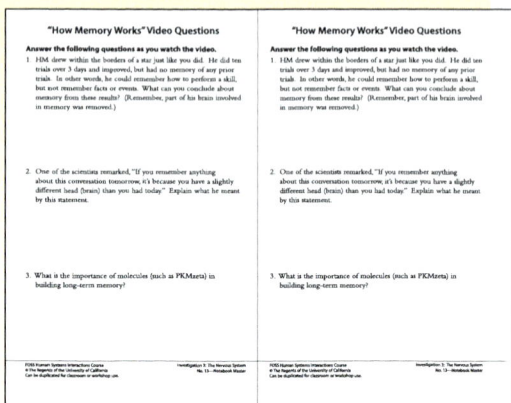

No. 13—Notebook Master

MATERIALS for
Part 4: Learning and Memory

Provided equipment
For each student
- 1 Mirror
- 1 Index card
- 1 *FOSS Science Resources: Human Systems Interactions*
- • "Memory and Your Brain"

Teacher-supplied items
For each student
- 1 Pencil
- 2 Pieces of scratch paper

For each group
- 1 Marking pen
- 1 Piece of chart paper

For the class
- 1 Stopwatch or timer
- • Chart paper (optional)
- 1 Marking pen
- • Materials for memory test
- 1 Book   1 Comb   1 Cork   1 Dollar bill
- 1 Fork   1 Key   1 Paper bag 1 Paper clip
- 1 Ruler   1 Stapler

FOSSweb resources
For each student
- • Notebook sheet 12, *Mirror Drawing*
- • Notebook sheet 13, *"How Memory Works" Video Questions*
- • *Investigation 3 I-Check*
- • *Posttest*

For the class
- • Video: *How Memory Works* (link to this video only)

216 Full Option Science System

Part 4: Learning and Memory

For the teacher
- Teacher master N, *Memory Set: Hear Only*
- Teacher master O, *Memory Set: Read Only*
- Teacher master P, *Memory Set: Hear and Read*
- Teacher master Q *Memory Set: Hear, Read, and Write*
- Teacher master R, *Memory Set: Hear, Read, Write, and See Object*
- *Embedded Assessment Notes*
- *Assessment Record*
- Teaching slides, 3.4

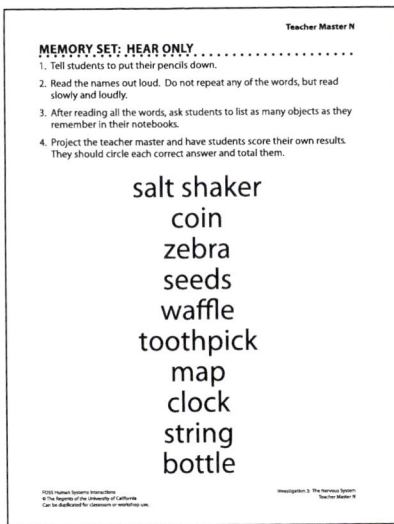
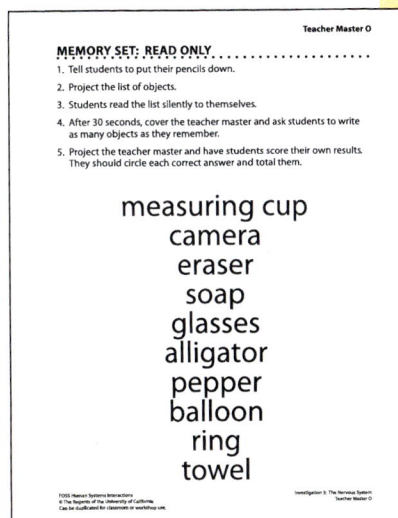

Teacher masters N–R

Human Systems Interactions Course—FOSS Next Generation   **217**

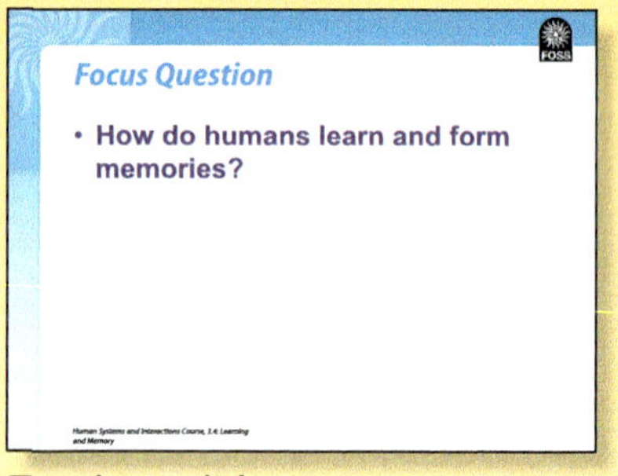

INVESTIGATION 3 – *The Nervous System*

GETTING READY for
Part 4: *Learning and Memory*

Quick Start

Schedule	3 active investigation sessions, including 1 reading 2 sessions for review and *I-Check 3* 2 sessions for review and *Posttest*
Preview	• Preview the FOSSweb Resources by Investigation for this part (such as printable masters, teaching slides, and multimedia) • Preview the reading: "Memory and Your Brain," Step 15 • Preview video: *How Memory Works*, Step 17
Print or Copy	**For each student** • Notebook sheets 12, 13 • *I-Check 3* • *Posttest* **For the teacher** • *Embedded Assessment Notes* • *Assessment Record*
Prepare Material	• Plan for mirror writing **A** • Gather materials for the memory list **B**
Plan for Assessment	• Review Step 20, "What to Look For" in the notebook entry • Plan for review and *I-Check 3*, Steps 21, 22 **C** • Plan for review and *Posttest*, Steps 23, 24 **D**

Focus Question
• How do humans learn and form memories?

Teaching slides, 3.4

218   Full Option Science System

Part 4: Learning and Memory

Preparation Details

A▶ Plan for mirror writing

Practice using a pencil, index card, and mirror to trace inside the boundary lines of one star on notebook sheet 12, *Mirror Drawing*. This will help you explain the process in Step 5.

If the mirrors are brand new, remove the plastic protective coating. You can also instruct the first students to do this.

B▶ Gather materials for the memory list

Gather one of each item for the final memory list in Step 13: book, comb, cork, dollar bill, fork, key, paper clip, ruler, stapler, and paper bag. Put these in a bag or under a piece of paper near the front of the room so that they stay hidden until you read the name of the object. Arrange them so that you can quickly pull each one out in the right order and then hide it again.

C▶ Plan for review and *I-Check 3*

If students will be taking the I-Check online, open the session on FOSSmap. If not, make copies of the I-Check for each student.

At least one day before taking the I-Check, allow time for students to review their notebook entries to prepare for the benchmark assessment. When taking the I-Check, students should not use their notebooks, but the notebooks are a good tool to use when students later reflect on their answers.

Use the coding guides in the Assessment chapter to score the items or review the FOSSmap reports that automatically score most items.

D▶ Plan for *Posttest*

Plan to have students review the key ideas from the entire course the day before students take the *Posttest* at the end of the course. If students will be taking the *Posttest* online, open the session on FOSSmap. If not, make copies of it for each student.

Students should work independently to answer the questions. A coding guide is available in the Assessment chapter.

Human Systems Interactions Course—FOSS Next Generation

INVESTIGATION 3 – The Nervous System

FOCUS QUESTION

How do humans learn and form memories?

GUIDING *the Investigation*
Part 4: *Learning and Memory*

SESSION 1

Students will . . .
- Think about learning and how memories form (Steps 1–3)
- Engage in mirror writing tasks as an example of learning process (Steps 4–7)
- Engage in metacognition about strategies used to learn a skill (Step 8)
- Share results about individual experiences (Steps 9–11)

SESSION 2

Students will . . .
- Review learning strategies (Step 12)
- Engage in memory tasks using one or more senses each time (Step 13)
- Share and analyze results and strategies used to remember different kinds of objects (Step 14)

SESSION 3–4

Students will . . .
- Read and discuss "Memory and Your Brain" article (Steps 15, 16)
- Gather information from video "How Memory Works" (Step 17)
- Review vocabulary (Step 18)
- Answer the focus question (Steps 19, 20)
- Review and discuss notebook entries and list big ideas (Step 21)

SESSIONS 5–7

Students will . . .
- Demonstrate understandings by responding to *Investigation 3 I-Check* (Step 22)
- Review *I-Check 3* and big ideas for course (Step 23)
- Demonstrate understandings by responding to the *Posttest* for the course (Step 24)

Part 4: Learning and Memory

SESSION 1 45–50 minutes

1. ## Introduce learning and memory
   Remind students that they have been studying how we interact with the environment, sensing and responding to stimuli. They have thought about how their brain processes sensory data, comparing new experiences to previous experiences. Now it is time to think about how our brains learn and form memories.

2. ## Focus question: How do humans learn and form memories?
   Pose the focus question, write or project it on the board, and have students write it in their notebooks.

   ➤ How do humans learn and form memories?

   Students do not need to respond to the focus question at this time but should leave the rest of the page blank.

   > **TEACHING NOTE**
   > This short activity is very engaging. Be prepared to share your own earliest memory in addition to having students share theirs.

3. ## Recall early memories
   Ask students to recall their earliest memory and share it with a partner. Call on a few students to share their memories with the class. Tell students,

   Learning is defined as the process of acquiring new knowledge and skills. Learning happens when your brain forms **memories**. Memory refers to the storage and recall of information, including past experiences, knowledge, skills, and thoughts.

4. ## Introduce mirror writing
   Tell students that they are going to learn something new and think about how they do that. Hold up a mirror and explain that they will be learning how to write by looking in a mirror. Warn them that the task may be difficult, and that the point is not to get it right immediately. Rather, they should think about how they approach the task and the strategies they use to learn a different new skill.

   > **TEACHING NOTE**
   > It is important that students focus on learning the skill of mirror writing, and not on getting it right immediately. This is a challenge!

5. ## Demonstrate the setup
   Hold up an index card and demonstrate how to poke a pencil through its center, thus installing a screen on the pencil. Demonstrate how to position the pencil with the index-card screen, the mirror, and notebook sheet 12, *Mirror Drawing,* so that students can see their drawing in the mirror, but cannot see their pencil point directly.

Human Systems Interactions Course—FOSS Next Generation

INVESTIGATION 3 — *The Nervous System*

6. Explain the tasks

Tell students,

Your first task is to write your name on your notebook sheet, so that it looks correct in the mirror. Use the mirror to guide and check your progress.

Then use the mirror to trace the shape of one star, staying inside the double lines. Use the top star only for now.

Explain that they must not look directly at their pencils, and that they should not erase anything, because they are trying to see if they get better as they practice tracing around the top star.

7. Begin mirror writing

Distribute a copy of notebook sheet 12, *Mirror Drawing*, to each student. Have Getters get mirrors and index cards, and let the name writing and drawing begin. Cruise the room, helping students coordinate the screened pencils and mirrors to work on the tasks. If students finish early, have them write messages to each other, using mirror writing.

8. Help students reflect on their thinking

Encourage students to think about the strategies they use to learn something new. This process is called **metacognition**, thinking about one's own thinking. Encourage them to write down a few ideas in their notebooks.

9. Share results

Ask students to talk in their groups and share what strategies they used to try to learn the skill. After a few minutes, have each group share one or two observations and strategies. List their ideas on the board or a piece of chart paper. Use the following questions to aid discussion:

➤ *What errors persisted despite your best efforts? Why?*

➤ *What sense helped you most—touch or vision? Why?*

➤ *How did you feel when you tried the mirror writing?*

➤ *Have you ever felt that way when trying to learn something else that was new? What was it?*

➤ *How might you improve?* [Practice.]

10. Try mirror writing again

Suggest that if students practiced mirror writing, they might get good at it. Ask students to trace inside the lines of the second star. Ask students to think about whether the task has become easier.

EL NOTE

Provide students who need support with a reflection prompt such as "I have changed my thinking about _____; I want to know more about _____."

Part 4: Learning and Memory

11. Clean up

Ask students to return the mirrors to the materials station. They can save the index cards in their notebooks for another effort or toss them in the recycling bin.

SESSION 2 45–50 minutes

12. Review learning strategies

Ask students to remember the strategies they used to learn how to mirror write. Have a few students share their ideas. Say,

Remember, learning happens when your brain forms memories. We all are different and may have a different preferred strategy when it comes to trying to remember events or facts.

Tell them that they are now going to explore memory—to find out how they remember things.

13. Introduce memory tasks

Explain that the memory tasks are not for grades and that nobody else will see them. Say,

These tasks are to help you learn how you remember best. Don't worry if you do things differently than your neighbor. These memory tasks are not for you to compare with anyone. You can use what you learn about yourself to help you remember better in this class, in other classes, and even outside of school.

Explain that they will try to remember and record ten different words or objects, using several ways to get information about the objects into their brains. Describe the five different tasks.

- You will hear the names of the objects (hear only, TM N).
- You will read the names of the objects (read silently only, TM O).
- You will hear and see the names of the objects (hear and read, TM P).
- You will hear, see, and write the names (hear, read, and write, TM Q).
- You will hear, see and write the names, and see the actual objects (hear, read, write, and see object, TM R).

▶ **NOTE**
Have the teacher masters in order and ready to go for Step 13.

In their notebooks, have students title a new page "Memory Tasks." They should put a subheading for each test. The first is "Hear Only." Distribute scratch paper to students for the last two memory tasks involving writing.

Human Systems Interactions Course—FOSS Next Generation

INVESTIGATION 3 – *The Nervous System*

SCIENCE AND ENGINEERING PRACTICES

Analyzing and interpreting data

Constructing explanations

CROSSCUTTING CONCEPTS

Patterns

Hear only	4
Read only	9
Hear and read	13
Hear, read, and write	8
Hear, read, write, and see	22

> **TEACHING NOTE**
>
> Ask if any students used their hands to make motions or do other things to help them remember objects in the lists.

> **TEACHING NOTE**
>
> Research is constantly adding to our understanding of how memories form. Students might be interested to know more about diseases such as Alzheimer's, which appears to erase the ability to form new memories.

Conduct the tasks one at a time using the appropriate teacher master as a guide. Conduct them all in the same way.

a. Present the words or objects in the manner of the particular test.

b. When reading out loud, do not repeat the word; speak clearly and slowly.

c. If students are writing as part of the test, have them use scratch paper, and remove the paper when they are ready to recall their list from memory.

d. For the final test, hold up each object as you read its name, then hide it from sight as you go on to the next object.

e. Give students 1 minute to write the names of as many objects as they remember in their notebooks under the heading for the test.

f. Have students circle their correct answers and total them to get their score.

14. Share results

Have students order the tasks by the number they remembered. This will help them determine which task best assisted memory. Tally the number of students who remembered best on each task on the board. If students did equally well on several of the tasks, they should raise their hands for all those tasks (so the total might not equal the number of students in the class). Typical results might look something like those in the sidebar.

Ask students to discuss in their groups what the class results suggest. They can also share which task they personally did the best on and why they think that happened. After a minute or so, call on the Reporters from each group to share what they discussed. Ask,

▶ *What do the results tell you?* [The more senses used to acquire information, the more efficient the formation of memory.]

▶ *Why do you think that is?* [You are using different parts of your brain for each sense, so if you are using two senses, more of your brain is engaged and you may have a better chance at remembering.]

▶ *What strategies did you use to remember the objects?*

Remind students that each person has a different way of remembering things, so that they should keep in mind the strategies that work best for them. Tell students to share with a partner how they might use knowledge of their personal memory strategies to change the way they learn in their classes or outside of school.

Part 4: Learning and Memory

SESSION 3 45–50 minutes

READING *in Science Resources*

15. Read "Memory and Your Brain"
The article "Memory and the Brain" will help students consider how their brain forms memories. Use the guide on the following pages to lead an in-class reading of the article. Have students set up their notebooks to take notes about new ideas or questions. At breakpoints indicated in the reading guide, stop the class for a minidiscussion.

16. Use a reading comprehension strategy
Point out the image of the brain. The parts of the brain that are most involved in memory formation are labeled. Give students a few moments to share what they know about the parts of the brain and their function. Suggest they draw a brain outline in their notebook and label and describe the parts as they read about them. They could trace the brain outline using page 104 of the *FOSS Science Resources* book, "Brain Map of Sensory Activity."

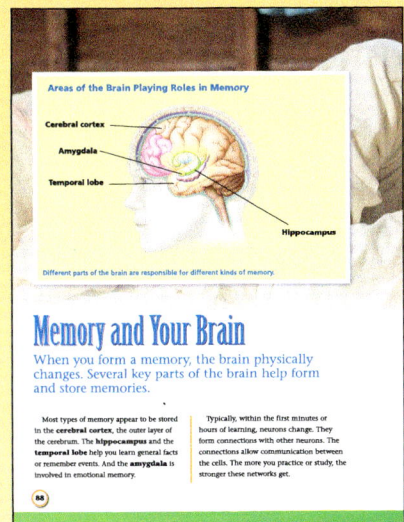

SCIENCE AND ENGINEERING PRACTICES

Obtaining, evaluating, and communicating information

ELA CONNECTION

This suggested strategy addresses the Common Core State Standards for ELA for literacy.

RST 7: Integrate quantitative or technical information expressed in words in a text with a version of that information expressed visually.

TEACHING NOTE

For more information on scaffolding literacy in FOSS, see the Science-Centered Language Development in Middle School chapter in Teacher Resources.

Human Systems Interactions Course—FOSS Next Generation

INVESTIGATION 3 – The Nervous System

Use academic language:
Read aloud or have a volunteer read the title and first two sentences. Pause to have students visualize what this might look like. Have them read the rest of the page independently and add information to their brain diagram. Before continuing, have students pair up and talk about what they've learned about neurons and how they think they are involved in memory and learning.

ELA CONNECTION

This suggested strategy addresses the Common Core State Standards for ELA for literacy.

L 6: Acquire and use academic and domain-specific words and phrases.

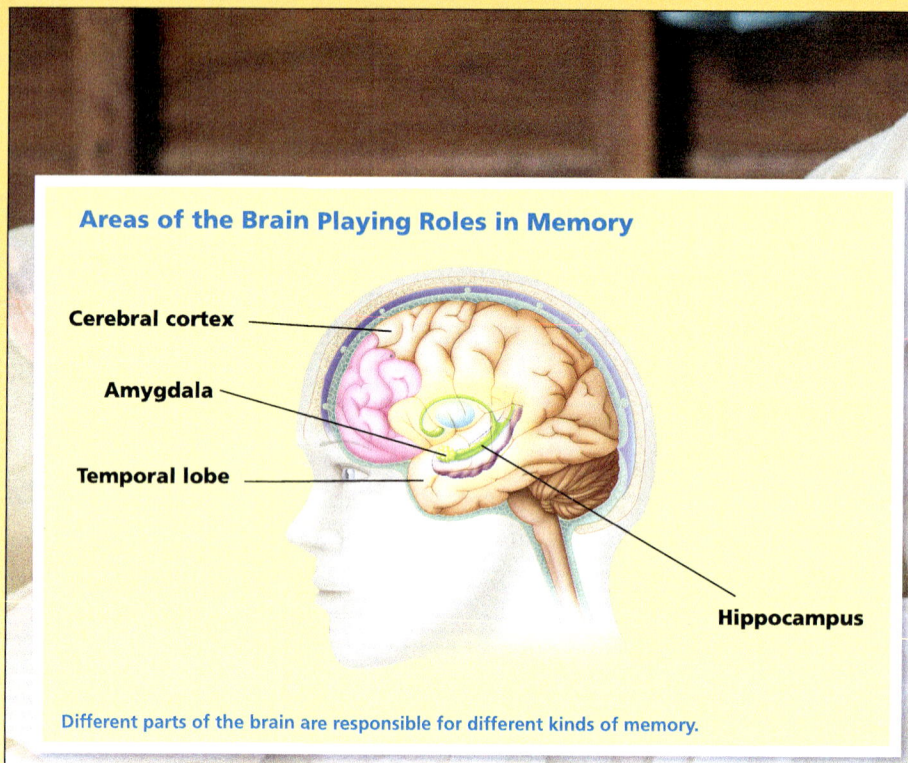

Areas of the Brain Playing Roles in Memory

- Cerebral cortex
- Amygdala
- Temporal lobe
- Hippocampus

Different parts of the brain are responsible for different kinds of memory.

Memory and Your Brain

When you form a memory, the brain physically changes. Several key parts of the brain help form and store memories.

 Most types of memory appear to be stored in the **cerebral cortex**, the outer layer of the cerebrum. The **hippocampus** and the **temporal lobe** help you learn general facts or remember events. And the **amygdala** is involved in emotional memory.

Typically, within the first minutes or hours of learning, neurons change. They form connections with other neurons. The connections allow communication between the cells. The more you practice or study, the stronger these networks get.

226 Full Option Science System

Part 4: Learning and Memory

The neural connections that form our memories are strengthened during sleep.

B

What Affects How Memories Form?

Four things affect how memories form. They are paying attention, getting enough sleep, emotions, and the number of senses.

Paying Attention

"Why don't you remember what I told you?"

"I wasn't paying attention."

That's the problem. Doing many things at once divides our attention. Much research has focused on media multitasking. It looks at students who use multiple kinds of technology at the same time. Guess what the studies are finding? These students make many mistakes in memory tests. They are too distracted to remember what happened. So cut out distractions when you are trying to learn new things.

Multitasking is not as efficient as it sounds. When your brain is constantly switching focus and responding to interruptions, you'll find it hard to concentrate when you need to.

B

Draw evidence from informational text: Have students read the rest of the article independently. Have them continue to take notes. Suggest they use the section titles as questions they can answer from the text. The first one is already in the form of a question. They should answer the question citing evidence from the text and include a personal response, (i.e, According to ___. This makes me think about ___.)

ELA CONNECTION

This suggested strategy addresses the Common Core State Standards for ELA for literacy.

WHST 9: Draw evidence from informational texts to support analysis, reflection, and research.

Investigation 3: *The Nervous System* 89

Human Systems Interactions Course—FOSS Next Generation 227

INVESTIGATION 3 – The Nervous System

Evaluate an argument:
Here students might ask, "How does getting enough sleep affect memory?" Suggest students identify the claim made by the study mentioned in the article and the evidence used to support that claim. They can also add any other evidence that would support or refute that claim based on their own experiences.

ELA CONNECTION

This suggested strategy addresses the Common Core State Standards for ELA for literacy.

SL 3: Delineate and evaluate a speaker's argument.

Getting Enough Sleep

A 2014 study investigated how sleep affects mouse neurons. When mice learn new things, spines appear on the dendrites of neurons they use. These spines can connect to other neurons. The learning or memory is stored. When mice sleep after the new tasks, those neurons grow more than if they don't sleep.

This study concludes that sleep contributes to storage of new memories. Your memory will work better if you get enough sleep. And that means you will do better on tomorrow's test if you get a good night's sleep.

In humans and other animals, sleep has a profound, positive effect on learning and memory.

Part 4: Learning and Memory

Emotions

The amygdala helps us recognize emotionally significant events. You may not often remember what you had for dinner a week ago. But say you heard some really bad news during dinner. You may remember the details of that event, including what you were eating. Positive emotions can also affect how well you remember an event, but negative emotions are even more powerful.

> **Take Note**
> Think back to when your teacher asked you to remember your oldest memory. What emotions were associated with the memory you recalled?

 Here students might ask, "What do emotions have to do with memory?" After students write down their personal connections, give them a few minutes to share them with a partner.

ELA CONNECTION

This suggested strategy addresses the Common Core State Standards for ELA for literacy.

SL 1: Engage in collaborative discussions.

INVESTIGATION 3 — The Nervous System

Discuss the reading:
At the end of the reading, provide time for students to respond to the questions from their own personal perspective. You might ask them "How does using more senses help learning?"

More Senses Help Learning

Remember the memory tasks you tried? Which ones were easiest for you? For most people, the more senses that are used at the same time, the more they can remember. By writing things down or saying them out loud, you can learn better than by simply reading. Different kinds of neural networks form. If you can't remember something you saw, maybe you can remember what you heard or wrote.

If too many senses are stimulated at the same time, however, it can be confusing. Your brain cannot process too much information at once.

Our brains are incredibly complex and always changing. As we learn and remember experiences, the brain builds networks of neurons that maintain that learning and memory. Those neural networks help make us who we are.

Think Questions

1. How do you best study for a test or learn new information?
2. How can you improve your memory? What strategies have you heard of besides the ones in this article?

Taking notes can help you remember what you see and hear in class.

92

ELA CONNECTION

This suggested strategy addresses the Common Core State Standards for ELA for literacy.

SL 6: Adapt speech to a variety of contexts and tasks.

230　　Full Option Science System

Part 4: Learning and Memory

17. View video

Give students a short introduction to the video, *How Memory Works*. Tell them,

This short video starts with the story of a man known by his initials, HM. HM suffered from a disease called epilepsy, which caused seizures, or brief spurts of uncoordinated brain activity, when he was 10 years old. As he grew older, the seizures became worse. He and doctors finally decided that the only way to stop the seizures was to remove part of his brain, including the hippocampus, the amygdala, and parts of the cerebral cortex.

➤ *What do you think happened to him?*

Distribute notebook sheet 13, *"How Memory Works" Video Questions*. Play the video, stopping at appropriate moments to clarify or to give the students time to answer the questions. Note the references to HM doing the mirror writing task.

Move into the final wrap-up after discussing the video.

NOTE: The study of memory formation in the brain is an exciting, continually changing field. The video was made in 2009, when PKMzeta was understood to be one of the key long-term memory molecules. However, a study released in 2013 suggests otherwise. Keep your eyes open for current scientific studies that reveal new understanding. Visit FOSSweb for links to current scientific studies about memory.

18. Review vocabulary

Give students a few moments to review the vocabulary developed in this part. Students should highlight the words and elaborate on their notes and diagrams in their notebooks. This is also a good time to update their vocabulary indexes if they haven't already done so.

19. Answer the focus question

Students should summarize what they've learned. Remind them to include the new vocabulary words in their response to the focus question.

➤ *How do humans learn and form memories?*

Tell students you will be looking at their entries to see how well they understand the concepts of learning and memory so they should try their best to explain their thinking as clearly and comprehensively as possible. Ask them to include a description of what they have discovered about their own learning through the memory tasks.

▶ **NOTE**
This 10:52 minute video from PBS can be accessed from the NOVA site: It starts at 39:24 on the longer video.

http://www.pbs.org/wgbh/nova/body/how-memory-works.html

learning
memory
metacognition

E L N O T E

Provide students who need support with writing frames:
"The question we wanted to answer was _____ .
To find out we explored _____ .
We learned about the different _____ .
For example, _____ .
What I found most interesting was _____ .
I still have questions about_____ ."

Human Systems Interactions Course—FOSS Next Generation

INVESTIGATION 3 – The Nervous System

SCIENCE AND ENGINEERING PRACTICES

Obtaining, evaluating, and communicating information

20. Assess progress: notebook entry

Have several students in each class turn in their notebooks open to the page with the focus question answer showing. Spend 15 minutes after class reviewing this selection of student work. Use *Embedded Assessment Notes* to record your findings. Plan next-step strategies as needed.

What to Look For

- *Students share that when neural networks are strengthened in the brain, memories form.*
- *Students describe their learning in the memory tasks as being something that was difficult; they needed to focus and pay attention; they needed to practice over and over to make it easier to do.*

Part 4: Learning and Memory

SESSION 4   45–50 minutes

WRAP-UP

21. Review notebook entries

Give groups time to look at their notebook entries for the investigation and generate a list of big ideas (including drawings or illustrations) on a piece of paper or chart paper. Each group should be prepared to explain what they have written on their chart. The charts can be posted around the room for the rest of the class session. Students can add a list of big ideas to their own notebooks, and you can record a class list.

Here are the big ideas that should come forward.

- Sensory receptors receive information from the environment and transmit that information to the brain.
- Mechanoreceptors respond to physical stimuli, chemoreceptors to chemical molecules, and photoreceptors to light.
- A neural pathway consists of a series of neurons connected by gaps called synapses.
- Electric signals travel along a neuron until they hit a synapse, where they cause chemical neurotransmitters to be released and move across the gap, allowing the message to continue.
- Long-term memories form as permanent neural networks are built and reinforced.

DISCIPLINARY CORE IDEAS

LS1.A: Structure and function
LS1.D: Information processing

SESSION 5   45–50 minutes

22. Assess progress: I-Check

When students are ready, administer *Investigation 3 I-Check*. You should administer the I-Check now, review the answers at the same time as your review for the Posttest. The answer sheets and coding guides are in the Assessment chapter.

If students finish early, they can begin to review their notebook entries for the big ideas in the entire course.

Human Systems Interactions Course—FOSS Next Generation

INVESTIGATION 3 — The Nervous System

 **SESSION 6** *45–50 minutes*

23. Review *I-Check 3* and prepare for *Posttest*
Based on your review of the I-Checks, plan for next-step strategies. On that same day, give groups time to look at their notebook entries for the entire course and generate a list of big ideas.

This is a good time to return their *Entry-Level Survey* to students (if you haven't done so already). Ask students to read what they wrote before instruction and to construct a response to the question now that they have learned about human systems interactions.

 **SESSION 7** *45–50 minutes*

24. Assess progress: *Posttest*
When students are ready, administer the *Posttest*. The answer sheets and coding guides are in the Assessment chapter.

234 Full Option Science System

Extending the Investigation

EXTENDING *the Investigation*

- **Measuring sound levels**
  Find out more about measuring sound levels in decibels. Explore phone apps that measure sound levels.

- **Research different kinds of hearing aids**
  In the "Hearing" article in the *FOSS Science Resources* book, there is a diagram of one type of hearing improvement device called a cochlear implant. There are other devices to improve hearing. Have students research the devices that have been designed to improve hearing and describe how they function.

- **Compare sensory systems of different animals**
  Students can select a type of organism and compare its sensory systems to that of humans or they could select one of the sensory systems in humans (vision, hearing, smell) and compare a similar system in a variety of organisms (vision, hearing, smell). They could also research sensory systems that other animals have that humans do not have.

- **Find out more about Helen Keller**
  Students can read a biography about Helen Keller, focusing on how she learned and communicated.

INVESTIGATION 3 – The Nervous System

HUMAN SYSTEMS INTERACTIONS — *Assessment*

THE FOSS ASSESSMENT SYSTEM *for Middle School*

The FOSS assessment system was developed through the NSF-funded ASK Project over a period of 5 years, partnering with more than 500 teachers and their students. The system developed includes both **formative** and **summative** assessments. The purpose of formative assessment is to look at students' thinking as it is developing during instruction. It provides evidence of progress, information to guide future instruction, and is generally diagnostic. The purpose of summative assessment is to measure achievement after instruction is completed, and is generally evaluative.

Embedded assessments. Assessing the three dimensions envisioned in the NRC *Framework* and the NGSS performance expectations requires you to peek over students' shoulders, listen to their conversations, and ask critical questions while they are in the act of doing science or engineering. Observing the rich conversation and actions among students and asking questions as they investigate phenomena or design solutions to problems provides important information about student progress and next instructional moves needed. (See more about performance assessment in the section "Embedded Asssessment.")

Notebook entries also serve as assessment opportunities to gather evidence of students' thinking. Research shows that regularly reviewing student work can significantly improve students' achievement. Use the reflective assessment practice to make this quick and easy. The goal is to look at student work on a daily basis—frequency is the key to improvement. (See more about this practice in the section "Embedded Assessment.")

Benchmark assessments. The **Entry-Level Survey** is administered before instruction begins. Use it to help you determine students' background knowledge—how are students thinking about the topic now; what emerging conceptions do they have that you will be able to build upon as you move through the course? Encourage students to answer the questions as best they can so you get the information you need to move instruction forward. Students will return to the Entry-Level Survey at the end of the course to look at how their thinking has changed and to review for the Posttest.

Contents

The FOSS Assessment System for Middle School	237
Assessment for the NGSS	239
Embedded Assessment	244
Benchmark Assessment	250
Next-Step Strategies	252
FOSSmap and Online Assessment	258
Entry-Level Survey	260
Investigations 1–2 I-Check	264
Investigation 3 I-Check	272
Posttest	280
Benchmark Assessment Summary Chart	288

▶ **NOTE**
Always check FOSSweb for updates to the benchmark assessments before beginning the course.

Full Option Science System

HUMAN SYSTEMS INTERACTIONS — *Assessment*

I–Checks, given at the end of investigations, serve as checkpoints for student learning. To track achievement (a summative use of I–Checks), use the coding guides in this chapter. To use I–Checks for formative assessment, return I–Checks to students unmarked. Research evidence suggests that students learn more when they are given time to reflect on, evaluate, and improve their responses. When students check their own understanding, you are creating a class culture of continuous improvement, using assessment as a tool in the service of learning.

Students take the **Posttest** after instruction is completed. This is the summative assessment for the course. You can also use the Posttest for formative instructional evaluation. Make notes about things you might want to focus on or do differently next time you teach the course.

See more about these assessments in the section "Benchmark Assessment."

Assessment for the NGSS

ASSESSMENT *for the NGSS*

A Framework for K–12 Science Education (National Research Council, 2012) and *Next Generation Science Standards* (National Academies Press, 2013) provides a new vision for what science education ought to be. It emphasizes the idea that science education should resemble the way that scientists work. Students plan and conduct investigations, gather data, and then construct explanations and engage in argumentation in order to build their understanding of how the natural world works. The NRC *Framework* also emphasizes instruction guided by thoughtful progressions of learning. In other words, students are expected to construct and discuss explanations and models of their understanding in more and more sophisticated ways as they work through the grades. Assessment plays an important role in this vision—assessment is the bridge between teaching and learning.

Developing Assessments for the Next Generation Science Standards was published by a committee of the National Research Council charged with making recommendations for assessment design that will best support the vision of science education elaborated in the NRC *Framework* and in *Next Generation Science Standards*. Several key points are described in the National Research Council assessment report. The six points relevant to classroom curriculum development are listed below with a brief description about how FOSS embodies each one.

Assessment developers need to take a rigorous approach to the process of designing and validating assessments. The FOSS assessment system is based on a construct modeling approach to assessment design. That means that we have done the homework needed to describe a conceptual framework that provides evidence of students' progressive learning (see the course Framework chapter) and we have done the technical work needed to ensure that assessment tasks provide valid evidence of students' learning. Using the FOSS assessment system, you can be sure that the assessments are valid and reliable. This is one of the reasons that we do not provide an item bank from which to choose items. Instead, FOSS provides integrated sets of items that have been statistically analyzed as a cluster of items that provide detailed evidence of student learning.

Assessment systems should include formative and summative tasks. FOSS includes both formative and summative assessments. The embedded assessments—those assessments that occur with each lesson (part)—were originally designed to be the formative-assessment component of the system. As we were developing the system, however, it became clear that although the I-Checks (most of the benchmark

> *"Assessment plays an important role in this vision—assessment is the bridge between teaching and learning."*

Human Systems Interactions Course—FOSS Next Generation   **239**

HUMAN SYSTEMS INTERACTIONS — *Assessment*

component) were meant to be summative, using them for formative purposes is even more powerful than if they are used only to produce a grade. When students take an active role in their learning, using these tools for self-assessment, achievement improves dramatically. You can find out more about how to use benchmark assessments for formative assessment in this chapter.

Assessment systems should support classroom instruction. The main purpose of the FOSS assessment system is to support classroom instruction—to provide the bridge between teaching and learning. Teachers need information daily about what students have learned or may be confused about. FOSS teachers spend only 15 minutes at the end of the day using the reflective-assessment practice to gather evidence that helps them determine the next instructional steps. Are students ready to move on to the next lesson, or do they need some additional clarification? Research from the ASK Project showed that implementation of this technique can make a significant difference in students' achievement. (See the next page for more about ASK.)

Assessment should provide multiple and varied assessment opportunities and be used to locate students along a sequence of progressively more complex understandings. FOSS embedded assessments provide opportunities for you to assess students' conceptual development, science and engineering practices, and crosscutting concepts to approach science investigation and engineering problems. The FOSS assessment design is built on a construct modeling process, which allows us to provide descriptive diagnostic reports for student progress. The Framework chapter has examples of the conceptual frameworks and progressions used to define how science concepts are developed from kindergarten through grade 8.

Assessment tasks should consist of multiple components in order to measure all three dimensions of the NRC *Framework* and NGSS. The FOSS assessment system provides multiple tools and strategies to assess the three dimensions: (1) science and engineering practices, (2) disciplinary core ideas, and (3) crosscutting concepts. For example, FOSS suggests using parts of investigations as opportunities for performance assessments. The performance-assessment tasks are very much like those provided in *Developing Assessments for the NGSS* (NRC, 2014). Students must work like a scientist to answer a question or solve a problem. Then you look at how students use crosscutting concepts to determine what practices they need to use. For example, if the question requires understanding of cause and effect, students set up an investigation in which they control variables. While items on assessment tools will often focus on a single dimension, the performance assessments provide the opportunity to assess more at one time.

"When students take an active role in their learning, using these tools for self-assessment, achievement improves dramatically."

Assessment for the NGSS

Assessment systems should include an interpretive system.
FOSS provides extensive support for you to interpret assessment information. For each embedded assessment, specific information in Guiding the Investigation section describes what to look for in student responses to assess progress. Coding guides are provided for each item on benchmark assessments (I-Checks and Posttest). If students take an assessment online, and then you can use FOSSmap to run a number of diagnostic and summary reports (delivered as PDFs or files to be loaded into spreadsheet programs). Some of these reports provide you with information about class progress; others provide individual students and their parents with information about what students know and what they still need to work on. (See the FOSSmap and Online Assessment section in this chapter.)

Developing Assessment for the NGSS (NRC, 2014, Page 190) states, "To monitor science learning and adequately cover the breadth and depth of the performance expectations in the NGSS, information from external on-demand assessments [state assessments] will need to be supplemented with information gathered from classroom-embedded assessments." If you are implementing the FOSS assessment system, you will be well on the way to meeting this requirement when state science assessment systems are enacted.

The FOSS assessment system was developed as part of the NSF-funded ASK Project over a period of 5 years, partnering with more than 500 teachers and their students. We know that teachers can employ this assessment system for the benefit of their students, and we know that students achieve more. Perhaps even more important is the change in classroom culture that occurs when assessment is thoughtfully employed as the bridge between teaching and learning. Assessment is no longer the black cloud hanging over the class. It encourages students to adopt a growth mind-set—if I continue to work hard, I can make progress. It models what scientists do. Scientists use the information they have to argue the best theories they can, but they keep an open mind so when new evidence emerges, they can incorporate it into their thinking. That's what good curriculum and assessment are all about, too.

> *"To monitor science learning and adequately cover the breadth and depth of the performance expectations in the NGSS, information from external on-demand assessments will need to be supplemented with information gathered from classroom-embedded assessments."*
> *(National Research Council, 2014)*

HUMAN SYSTEMS INTERACTIONS — *Assessment*

NGSS Performance Expectations

"The NGSS are standards or goals, that reflect what a student should know and be able to do; they do not dictate the manner or methods by which the standards are taught.... Curriculum and assessment must be developed in a way that builds students' knowledge and ability toward the PEs [performance expectations]" (*Next Generation Science Standards*, 2013, page xiv). The FOSS assessment system includes embedded, performance, and benchmark assessments. The chart displayed here shows where each performance expectation is assessed in the middle school life science courses. These assessments help students build knowledge and ability in concert with the active investigations and readings to meet the goals of the NGSS.

Middle School Life Science NGSS Performance Expectations	FOSS Middle School Course
MS-LS1-1. Conduct an investigation to provide evidence that living things are made of cells; either one cell or many different numbers and types of cells.	Diversity of Life (Human Systems Interactions)
MS-LS1-2. Develop and use a model to describe the function of a cell as a whole and ways the parts of cells contribute to the function.	Diversity of Life
MS-LS1-3. Use argument supported by evidence for how the body is a system of interacting sub-systems composed of groups of cells.	Diversity of Life Human Systems Interactions
MS-LS1-4. Use argument based on empirical evidence and scientific reasoning to support an explanation for how characteristic animal behaviors and specialized plant structures affect the probability of successful reproduction of animals and plants, respectively.	Diversity of Life
MS-LS1-5. Construct a scientific explanation based on evidence for how environmental and genetic factors influence the growth of organisms.	Diversity of Life
MS-LS1-6. Construct a scientific explanation based on evidence for the role of photosynthesis in the cycling of matter and flow of energy into and out of organisms.	Populations and Ecosystems (Diversity of Life)
MS-LS1-7. Develop a model to describe how food is rearranged through chemical reactions forming new molecules that support growth and/or release energy as this matter moves through an organism.	Human Systems Interactions Populations and Ecosystems (Diversity of Life)
MS-LS1-8. Gather and synthesize information that sensory receptors respond to stimuli by sending messages to the brain for immediate behavior or storage as memories.	Human Systems Interactions

Note: Courses in parentheses incorporate these PEs, but they are not the main focus of the course.

Assessment for the NGSS

Middle School Life Science NGSS Performance Expectations	FOSS Middle School Course
MS-LS2-1. Analyze and interpret data to provide evidence for the effects of resource availability on organisms and populations of organisms in an ecosystem.	Populations and Ecosystems
MS-LS2-2. Construct an explanation that predicts patterns of interactions among organisms across multiple ecosystems.	Populations and Ecosystems
MS-LS2-3. Develop a model to describe the cycling of matter and flow of energy among living and non-living parts of an ecosystem.	Populations and Ecosystems
MS-LS2-4. Construct an argument supported by empirical evidence that changes to physical or biological components of an ecosystem affect populations.	Populations and Ecosystems
MS-LS2-5. Evaluate competing design solutions for maintaining biodiversity and ecosystem services.	Populations and Ecosystems Earth History
MS-LS3-1. Develop and use a model to describe why structural changes to genes (mutations) located on chromosomes may affect proteins and may result in harmful, beneficial, or neutral effects to the structure and function of an organism.	Heredity and Adaptation
MS-LS3-2. Develop and use a model to describe why asexual reproduction results in offspring with identical genetic information and sexual reproduction results in offspring with genetic variation.	Heredity and Adaptation (Diversity of Life)
MS-LS4-1. Analyze and interpret data for patterns in the fossil record that document the existence, diversity, extinction, and change of life forms throughout the history of life on Earth under the assumption that natural laws operate today as in the past.	Earth History Heredity and Adaptation
MS-LS4-2. Apply scientific ideas to construct an explanation for the anatomical similarities and differences among modern organisms and between modern and fossil organisms to infer evolutionary relationships.	Heredity and Adaptation
MS-LS4-3. Analyze displays of pictorial data to compare patterns of similarities in embryological development across multiple species to identify relationships not evident in the fully formed anatomy.	Heredity and Adaptation
MS-LS4-4. Construct an explanation based on evidence that describes how genetic variations of traits in a population increase some individuals' probability of surviving and reproducing in a specific environment.	Heredity and Adaptation
MS-LS4-5. Gather and synthesize information about technologies that have changed the way humans influence the inheritance of desired traits in organisms.	Heredity and Adaptation
MS-LS4-6. Use mathematical representations to support explanations of how natural selection may lead to increases and decreases of specific traits in populations over time.	Heredity and Adaptation

HUMAN SYSTEMS INTERACTIONS — Assessment

> **TEACHING NOTE**
>
> FOSS recommends that you do not grade notebook entries. This ensures a risk-free environment for students to write freely, knowing mistakes are part of learning. If you need to give a grade, have students complete a derivative product based on a notebook entry. Students might rewrite a focus-question answer, write up part of a lab, or revise a response sheet and turn it in, knowing that this product will be for a grade.

EMBEDDED ASSESSMENT

In FOSS middle school, the unit of instruction is the course—a sequence of conceptually related learning experiences that leads to a set of learning outcomes. A science notebook gives students a place to record their thinking and develop deeper understanding of the course content by articulating relationships, patterns, and conclusions, as well as by asking questions that will guide further exploration. Science notebook entries give both you and your students opportunities to review and reflect on students' thinking.

From the assessment point of view, a science notebook is a collection of student-generated artifacts that exhibit student learning. You can informally assess student skills, such as the ability to use charts to record data, while students are working with materials. At other times, you collect the notebooks and review them for insights or errors in conceptual understanding. The displays of data and analytical work provide a measure of the quality and quantity of student learning.

As you progress through the course, you will see different strategies used throughout the *Investigations Guide*. These will be marked with the notebook or assessment icon. As you try these strategies, take note of the positive effect that keeping notebooks have on students' work, as students continually practice expressing their conceptual development in writing. Embedded assessments help you better understand and address students' misconceptions.

Assessment Opportunities

Notebook entries serve as assessment opportunities for learning. Each part of each investigation is driven by a **focus question**. Each part usually concludes with students writing or revising an answer to the focus question in their notebooks. Their answers reveal how well they have made sense of the investigation and whether they have focused on the relevant actions and discussions.

At times, students use prepared **notebook sheets** to help organize and think about data. You can note how carefully students are making and organizing observations and how they think about analyzing and interpreting the data. Sometimes students answer a specific question that provides additional insight into understanding. You will find answers for notebook sheets in the Notebook Answers chapter.

Response sheets provide more formal embedded-assessment data. These are a specific kind of notebook sheet that assesses specific scientific knowledge that students often struggle with, giving you an

Embedded Assessment

additional opportunity to help students untangle concepts that they may be overgeneralizing or have difficulty differentiating.

Students also generate **free-form notebook entries** that can be used for assessing progress. These may occur when you choose to have students organize their own data, or when events in the classroom suggest a new aspect of students' learning that you want to know more about.

A **quick write** (or quick draw) is a question that students answer on a separate sheet of paper before instruction so you can analyze their prior knowledge and misconceptions. Collect quick writes and quickly sample them for insight into what students think about certain phenomena before you begin formal instruction. Knowing students' intuitive ideas (or prior knowledge) will help you know what parts of the investigations need the most attention. Make sure students date their entries for later reference. Quick writes can be done on a quarter sheet of paper or an index card. You collect them, review them, and return them to students to affix into their notebooks for self-assessment later in the investigation.

Performance assessments occur at times in the course as a way to check students' three-dimensional progress, checking science and engineering practices, crosscutting concepts, and disciplinary core ideas. These assessments happen during class as you circulate among student groups during their investigations. Sometimes you will simply watch what students are doing; at other times prompts or interview questions will be suggested.

Time Management

In order to collect enough data from embedded assessment to adequately inform instruction, plan to spend 15 minutes on each day of instruction, reviewing student learning by examining student work. In middle school, you face the challenge of having a large number of students. This may mean collecting only a portion of students' notebooks at a time to keep your workload manageable. A sample of student notebooks across your classes should represent the general levels of conceptual understanding that students have. Some work, such as quick writes and notebook sheets, can first be completed on separate sheets of paper. These are easier to collect, read, and later return to glue into students' notebooks.

HUMAN SYSTEMS INTERACTIONS — *Assessment*

Planning for Embedded Assessment

Embedded assessments are suggested for each investigation part. The Getting Ready Quick Start tells you what copies to make and how to plan for assessment. Here is an example from the Getting Ready Quick Start chart for Investigation 1, Part 1.

Print or Copy	**For each student** • *Entry-Level Survey* **For the teacher** • *Embedded Assessment Notes*

Plan for Assessment	• Review Step 12, "What to Look For" in the notebook entry

As you progress through the lesson, you will find a step in the **Guiding the Investigation** section that tells you what to assess. It will also give you a list of what to look for, including how students should be building and communicating their scientific knowledge, science and engineering practices, and crosscutting concepts as appropriate for the task.

> ### 12. Assess progress: notebook entry
> After students complete their responses to the focus question, collect a sample of notebooks from each class, and assess students' progress. The sample you select should give you a snapshot of the range of student understanding at this point in time. Ask students to turn in their notebooks open to the page you will be looking at.
>
> #### What to Look For
> - *The human body structures can be described at different scales or levels of organization that reflect increasing complexity.*
> - *Students list the levels in order of scale and complexity: atoms, molecules, cell structures, cells, tissues, organs, organ systems.*
>
> Plan to spend 15 minutes reviewing the selected sample of student responses. Use *Embedded Assessment Notes* to record your observations.

246 Full Option Science System

Embedded Assessment

Reflective Assessment Practice

Research shows that if you spend time reviewing student work each day, students achieve significantly more. Use the technique described here to make this a quick and easy process. If you have several classes, choose a random sample across the classes and plan 15 minutes to review student work. The important thing is that you are looking at student work as frequently as possible.

Make copies of *Embedded Assessment Notes* to record observations and assess students' work for all embedded assessments except for the performance assessments. Instructions for recording performance assessment progress begin on the next page.

Anticipate. Before class, fill in the investigation and part numbers and the date on a copy of *Embedded Assessment Notes*. Check the assessment step in Guiding the Investigation, and write in the concept(s) that you want to focus on. Limit your assessment to one or two important ideas.

Reflective-Assessment Practice

1. Anticipate
Use the *Investigations Guide* to plan for each part and determine embedded assessment.

2. Teach
Use Guiding the Investigation to teach the lesson. Collect student notebooks.

3. Review
Review students' work (15 minutes). Use "What to Look For" in Guiding the Investigation.

4. Reflect
Note trends and patterns you see in student understanding.

5. Adjust
Plan next instructional steps based on assessment reflection. Make notes for next year.

AFTER EACH PART

Human Systems Interactions Course—FOSS Next Generation

HUMAN SYSTEMS INTERACTIONS — Assessment

> **TEACHING NOTE**
>
> The intention here is to gather information that will help you know how to move the class forward in their conceptual development. Refrain from correcting students at this time, but plan to incorporate this information into your instruction to guide students as the class moves through the investigation. Notebook entries should reflect students' thinking, not the teacher's.

Teach the lesson. Keep in mind the "What to Look For" list as you teach the investigation. During the investigation, you might circulate among students as they work, looking over their shoulders and listening to their conversations. Use this information to plan strategies for feedback and discussion as the lesson continues.

Review. At the end of class, have students hand in their notebooks *open to the page that you will be reviewing*. (This will save you a lot of time.) Collect a sample of student notebooks from each of your classes throughout the day. Try to represent students of all abilities within the sample, so you can fairly judge whether your classes are mastering the concepts. Check your copy of *Embedded Assessment Notes* where you noted the one or two important ideas that you are looking for as you review student work. Make a tally mark for each student who gets it or doesn't get it. Consider students' developing conceptions and record patterns you observe to build on when reflecting and planning next steps.

Reflect. Take 5 minutes to summarize the trends and patterns (highlights and challenges) you saw in the "Reflections/Next Steps" section of *Embedded Assessment Notes*.

Adjust. Describe the next steps you plan to take in the next lesson. This is the key to formative assessment. You must take some action to help students improve their understanding. If you do this process frequently, the next steps should take only a few minutes of class time when the next part begins. A number of next-step strategies are listed later in this chapter.

Performance Assessments

FOSS has designated two opportunities for performance assessment in this course. Performance assessments focus on evidence of progress that students are able to work in the three dimensions woven together in an investigation: students' conceptual understanding, their science and engineering practices, and how they employ crosscutting concepts. Your main activity will be careful observation, but at times, you might step in with a 30-second interview to ask a few carefully crafted questions to learn more about students' thinking.

Record your observations on a copy of the *Performance Assessment Checklist*. There is a specific checklist for each performance assessment. The checklists are provided in two formats, one for making notes about groups, and the other for making notes about individual students. These are also available on FOSSweb as downloadable spreadsheets.

Embedded Assessment

The science and engineering practices, crosscutting concepts, and disciplinary core ideas that are assessed in the **Human Systems Interactions Course** during performances assessments are starred in the list below. For a more detailed look at these dimensions for middle school, see the Framwork Chapter.

Science and Engineering Practices

Asking questions and defining problems

Developing and using models ★

Planning and carrying out investigations

Analyzing and interpreting data ★

Using mathematics and computational thinking

Constructing explanations and designing solutions ★

Engaging in argument from evidence ★

Obtaining, evaluating, and communicating information ★

Crosscutting Concepts

Patterns

Cause and effect ★

Scale, proportion, and quantity

Systems and system models ★

Energy and matter ★

Structure and function

Stability and change

Disciplinary Core Ideas

LS1.A Structure and function ★

LS1.C Organization for matter and energy flow in organisms ★

LS1.D Information processing

PS3.D Energy in chemical processes and everyday life

> **TEACHING NOTE**
>
> "Engaging in the practices of science helps students understand how scientific knowledge develops; such direct involvement gives them an appreciation of the wide range of approaches that are used to investigate, model, and explain the world" (NRC Framework, page 42).

> **TEACHING NOTE**
>
> Crosscutting concepts help provide students with an organizational framework for connecting knowledge from the various disciplines into a coherent and scientifically based view of the world.

Human Systems Interactions Course—FOSS Next Generation

HUMAN SYSTEMS INTERACTIONS — *Assessment*

> **NOTE**
> Always check FOSSweb for updates to the benchmark assessments before beginning the course.

TEACHING NOTE

If you decide to return paper Entry-Level Surveys to the students to study for the Posttests, have them reflect on their original answers, then keep the answers they had or edit and add to them based on their new learning. It is important for students to know how their thinking changes based on new experiences and evidence.

BENCHMARK ASSESSMENT

Entry-Level Survey

The **Entry-Level Survey** is administered before instruction begins. Students are often uneasy about taking a "test" when they haven't yet had the instruction they need in order to do well. Help students view this survey as a tool for you to find out what they know before you begin instruction. It is important for them to know that the survey will not be graded, but is a helpful way for you to get to know what they learned in earlier grades so you can build their knowledge accordingly.

When you administer the survey, encourage students to answer the questions as best they can. Even if students think they don't know the answers, they should try to think about something related that they do know and apply that knowledge. Collect completed surveys, pull a random sample of students' responses from each of your classes. Review as many as you can for diagnostic purposes, but don't make any marks on them. You will find suggestions for what to look for later in this chapter. Plan to use the surveys to help students review for the Posttest at the end of the course.

I-Checks

At the end of most investigations, you will find a benchmark assessment called an I-Check. In some cases, content from two investigations is combined into one I-Check. This assessment serves as a checkpoint for student learning. To track achievement (a summative use of the I-Checks), use the coding guides in this chapter to code the items. We recommend that you score one item at a time, that is, code item 1 for all students, then move on to item 2, and so on. Even though you have to shuffle papers more, you will find it actually takes less time overall to code the assessments. Coding tends to be more consistent across students when you use this method, and it allows you to think about the performance of the whole class for each item and the kind of next steps you might take to correct learning weaknesses.

I-Checks can also be used for formative assessment. Research has shown that students learn more when they take part in reflection upon and evaluating their own responses. To do this, code students' tests but *do not write on them.* Learning tends to come to a halt when students see a grade on their work, especially if it is a poor grade. Instead, record marks in a grade book or computer spreadsheet. Use class discussion and **next-step strategies** to help students reflect on and refine their thinking. A good place to begin this process is by looking through the library of next-step strategies in the next section.

Benchmark Assessment

Choose three or four items from the I-Check that you want to discuss with students during class time. Determine the next steps you want to take. In the next session, start with these three or four items, using the strategies you chose. Then quickly go over other items, so that students can check the rest of their answers. When students check their own understanding, you are creating a class culture of assessment as a tool in the service of learning.

Here is a summary for using I-Checks.

1. Have students complete an I-Check unassisted.

2. Code I-Checks item by item, but do not make any marks on students' tests. Record codes in a grading program, spreadsheet, grade book, or on a copy of the *Assessment Record* specifically set up for that assessment. You can also find these on FOSSweb and can download them into a spreadsheet program. As you finish coding each item, note important points to review with students.

3. Return I-Checks to students. Help students reflect on and clarify their thinking. Use the next-steps/self-assessment strategies in this chapter to focus on the three or four items you chose for indepth reflection.

Posttest

Have students take the **Posttest** after all the investigations are completed. It can be administered in any of the ways described for the other benchmark assessments. Record student codes on a copy of the *Assessment Record* for the Posttest, or the downloadable spreadsheet.

Use the Posttests for formative evaluation of the course. Make notes about things you might want to focus on or do differently the next time you teach the course.

> **NOTE**
> Should you give assessments online or offline?
>
> You can make copies of assessment items using the assessment masters, or students can take the benchmark assessments online through the **FOSSmap** program. If you use FOSSmap, all items are coded by the program except open-response items. You can also run a variety of reports for yourself and for students.
>
> If you plan to use self-assessment activities, you can run the FOSSmap *Student Responses* report for each student, then project the items for students to reflect on. If you make copies of the assessment masters and have students take the tests offline, students have more room for corrections and additional notes as well as a permanent record of the assessment items.

Human Systems Interactions Course—FOSS Next Generation

NEXT-STEP STRATEGIES

The ASK Project (Assessing Science Knowledge) was funded by the National Science Foundation in 2003. For 5 years, the FOSS development team worked with nine districts/centers using the FOSS Program around the United States, including more than 500 teachers and their students as research partners. Based on the evidence provided by the assessments, we learned very quickly that assessment is worth doing only if follow-up action is taken to enhance understanding. Self-assessment provides students the opportunity to be responsible for their own learning and is a very effective tool for building students' scientific knowledge and practice.

Self-assessment is more than reading correct answers to the class and having students mark whether they got the right answer. Self-assessment must provide an opportunity for students to reflect on their current thinking and judge whether that thinking needs to change. This kind of reflective process also helps students develop a better understanding of what is expected in terms of well-constructed responses.

The Importance of Feedback

Teacher feedback is important for students' understanding of how their conceptual thinking can advance. The science notebook provides an excellent medium for providing feedback to individual students or the class regarding their work. Productive feedback calls for students to listen to or read a teacher comment, think about the issue raised by the comment, and act on it. The feedback might ask for clarification, an example, additional information, precise vocabulary, or review of previous work in the notebook. In this way, you can determine whether the problems with the student work relate to flawed understanding of the science content or a breakdown in communication.

Your feedback will also encourage students to take the notebook more seriously and to write more clearly as they attempt to create a complete record of the course. By writing explanations, students clarify what they know—and expose what they don't. And when students use their notebooks as an integral part of their science studies, they think critically about their thinking.

Self-assessment with benchmark assessment items requires deep, thoughtful engagement with complex ideas. It involves students in whole-class or small-group discussion, followed by critical analysis of their own work. For this reason, we suggest that you focus your

Next-Step Strategies

probing discussions on three or four questions from an I-Check, rather than on the entire assessment. The techniques described here are meant to give you a few strategies for entering the process of self-assessment. There is no single right way to engage students in this process: it works best when you change the process from time to time to keep it fresh. The strategies listed here are sorted into two groups: strategies for whole-class feedback, and strategies for individual-student and small-group feedback.

Strategies for the Whole Class

Key points. Discuss the item in question. After it is clear that students understand what is intended by the item prompt, call on individuals or groups to suggest key points that should be included in a complete answer. Write the key points on the board as phrases or individual words that will scaffold students' revision, rather than complete sentences they might mindlessly copy. When students return to their responses, they can number each of the key points they originally included in their answers, then add anything they missed.

Multiple-choice discussions. Students sit in groups of three or four, depending on how many possible answers there were for a given question. You assign an answer to each student (not necessarily the answer they chose). Each student is responsible for explaining to the group whether the assigned answer is correct and why or why not.

Class debate. A student volunteers an answer to an item on an assessment (usually one that many students are having trouble with, or one that elicits a persistent misconception). That student is in charge of the debate. He or she puts forth an answer or explanation. Other students agree or disagree, but must provide evidence to back up their thinking. Students are allowed to disagree with themselves if they hear an argument during the discussion that leads them to change their thinking. You can ask questions to keep the discussion on track, but otherwise you should stay on the sidelines.

Human Systems Interactions Course—FOSS Next Generation 253

HUMAN SYSTEMS INTERACTIONS — *Assessment*

Multiple-choice corners. When the class is equally split on what students have chosen as the correct response or only a few students have gotten the correct answer, have them meet in different corners of the room. Those who chose *A* go to one corner, those who chose *B* go to another corner, and so on. Each corner group needs to come up with an argument to convince the other corners that their answer is correct. As in a class debate, students are allowed to disagree with themselves if they become convinced their position is flawed or the reasoning of another group is more convincing. They then move to that corner and continue by helping their new group shape their arguments. (Don't be surprised if you find all students migrating to one corner before the presentation of arguments even begins!)

Revision with color. Another way that students can revise their answers after a key-points discussion is to use colored pens or pencils and the three Cs. As they read over their responses, they *confirm* correct information by underlining with a green pen; *complete* their responses by adding information that was missing, using a blue pen; and *correct* wrong information, using a red pen.

Review and critique anonymous student work. Use examples of student work from another class, or fabricate student work samples that emulate the problems that students in your class are having. Project the work using an overhead projector, document camera, or interactive whiteboard. Have students discuss the strengths and weaknesses of the responses. This is a good strategy to use when first getting students to write in their notebooks. It helps them understand expectations about what and how much to write.

Line of learning. Many teachers have students use a line of learning to show how their thinking has changed. When students return to original work (embedded or benchmark) to revise their understanding of a concept, they start by drawing a line of learning under the original writing. The line delineates students' original, individual thinking from their thinking after a class or group discussion has helped them reconsider and revise their thoughts.

Group consensus/whiteboards. Have students in each group (or pairs in each group) work together to compare their answers on selected I-Check questions during a class review. First, they create a response that the group agrees is the best answer. Groups write their responses on a whiteboard. When you give a signal, everyone holds their whiteboard up and compares answers. The class discusses any discrepant answers.

> **TEACHING NOTE**
>
> You can make inexpensive whiteboards using card stock and plastic sheet protectors. Students use whiteboard marking pens to write answers. Washcloths make great erasers.

Next-Step Strategies

Critical competitor. Use the critical-competitor strategy when you want students to attend to a specific detail. You need to present students with two things that are similar in all but one or two aspects. You can use any medium: two drawings, two pieces of writing, or a combination (such as a diagram compared to a description). The point is to compare two pieces of communication or representations in some way that will help students focus on an important detail they may be missing.

Sentence frames. After completing other self-assessment activities, have students consider all the items on the assessment and write a short reflection using sentence frames. This strategy allows students to choose one or two items that they would like to tell you more about.

> I used to think ____, but now I think ____.
>
> I should have gotten this one right, I just ____.
>
> I know ____, but I'm still not sure about ____.
>
> The most important thing to remember about ____ is ____.
>
> Can you help me with ____?
>
> I shouldn't have gotten this one wrong, because I know ____.
>
> I'm still confused about ____.
>
> Next time, I will remember to ____.
>
> Now I know ____.

HUMAN SYSTEMS INTERACTIONS — *Assessment*

Strategies for Individual Students and Small Groups

Self-stick note feedback. As you read through students' notebooks, add self-stick notes with comments or questions that help guide students to reflect on and improve their understanding.

Conferences. Use opportunities when students work independently, to confer with small groups or individual students.

Work in pairs. Have students work in pairs to continue to explore their ideas and refine their thinking. You might try to pair students so that a student who understands the concept well works with another student who needs some help.

Response log. Set up a response log at the back of students' science notebooks. Fold a notebook page in half, or draw a line down the center of the page. Write "Teacher Feedback" at the top of the left side of the page and "Student Comments" at the top of the right side of the page. When you want a student to think about something in his or her notebook, write your note in the "Teacher Feedback" column (or students can move a self-stick note from another page to the response log). Students then respond in the right-hand column, either addressing your comment there, or telling you which page to turn to in order to see how they have responded.

Next-Step Strategies

Response Log

Teacher Feedback | **Student Comments**

Response Sheet - Inv. 2
1. You said food and oxygen are carried in "blood tubes." What do you mean?

1. Food molecules are picked up by capillaries from the small intestine. Oxygen is picked up from the alveoli by capillaries.

Response Sheet - Inv. 3
1. Does the student's notes need any correction?

1. Yes, I forgot. There is no such thing as neurotransformers. They should be neurotransmitters.

Human Systems Interactions Course—FOSS Next Generation

HUMAN SYSTEMS INTERACTIONS — Assessment

FOSSMAP *and Online Assessment*

FOSSmap (fossmap.com) is the assessment management program designed specifically for teachers using the FOSS Program in middle school. This user-friendly system allows you to open online assessments for students, to review codes for student responses, and to run reports to help you assess student learning. FOSSmap was developed at the Lawrence Hall of Science in conjunction with the Berkeley Evaluation and Assessment Research (BEAR) center at the University of California, as part of a 5-year research and development project funded by the National Science Foundation. It is based on the tools developed in the Assessing Science Knowledge (ASK) project.

Embedded-assessment data can be entered into FOSSmap to provide evidence of differentiated instruction, to run reports for formative analysis, and to print notes to provide feedback in student notebooks. It is also a tool for teacher reflection and instructional improvement from year to year.

FOSSmap allows you to give students access to the **online assessment** system (fossmap.com/icheck). Students log in to this system to take the benchmark assessments (I-Checks and Posttest). Responses are automatically sent to the FOSSmap teacher program, where most are automatically coded. You will need to check short answers (mainly for correct answers that include inventive spelling), and to code open-response items. Students can answer open-response items on the computer or using paper and pencil, depending on the resources you have available.

If you choose to have students take the Entry-Level Survey on FOSSmap, the answers will not be coded, but you will be able to look at all of the students' responses in one convenient place, and make notes about each item for use when you teach the different parts of the course.

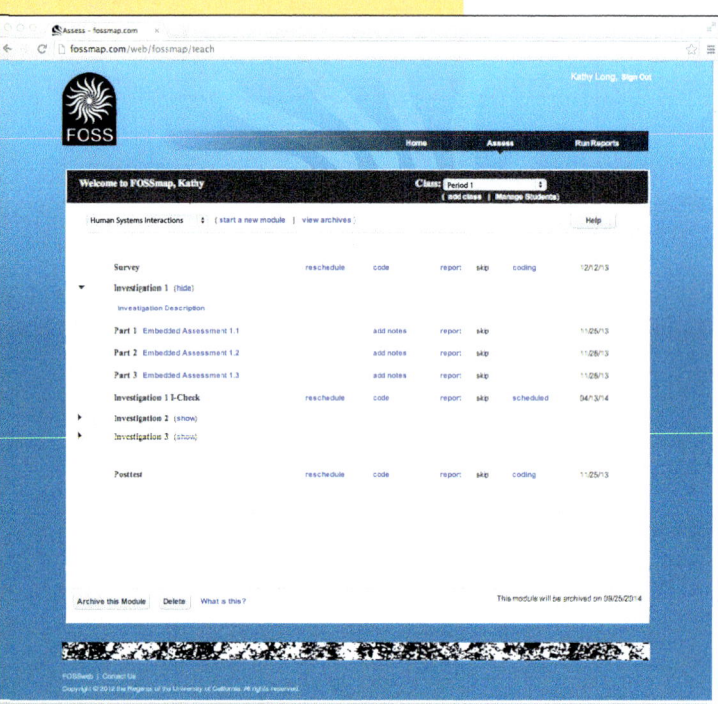

Navigation page and Embedded Assessment report

258   Full Option Science System

FOSSmap and Online Assessment

FOSSmap Reports

The **Code Frequency Report** tells you at a glance which items were problems for the class. Each bar on the report represents how many students received a particular code. The colored bars indicate how many students received the highest (max) code possible for the item. Green bars indicate that 70% or more students got the highest code. Yellow bars indicate that 51% to 69% of students got the highest code. Red bars indicate that 50% or fewer of the students got the highest code on that item. So the quick and easy way to use this report is to look for the red bars. The red-bar items are the ones you want to take back to students for self-assessment activities.

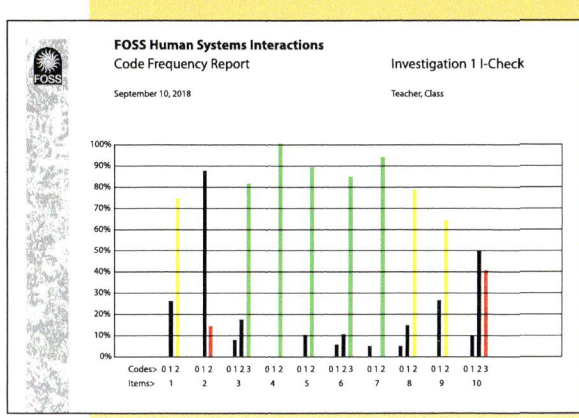

Run the **Class by Item Report** to get the details on each item, especially the "red bar" problem items from the Code Frequency Report. This report displays students' names for each response, with a brief description of what each code means in terms of full or partial understanding. The report helps you decide what steps need to be taken next.

The **Class All Codes Report** is essentially a spreadsheet that displays all of the codes that each student received for each item on the assessment. It also lists the max codes for each item, the cumulative total possible on the test, and the raw total for each student. You download the report from FOSSmap and you open it in any spreadsheet program.

The **Student Responses Report** provides a printout of individual students' responses to all items answered online (including open-response items if they were typed into the system). This report is useful for student self-assessment activities. You can project the items for class discussion, and students can make notes in their notebooks, add to, or revise answers based on the discussions during the self-assessment activities.

The **Student by Item Report** (a good report to send home to parents) lists all the items on a test and shows how individual students responded to each item. It also provides the correct answer, or max code, and a description of what the student knows or needs to work on, based on the evidence inferred from each item.

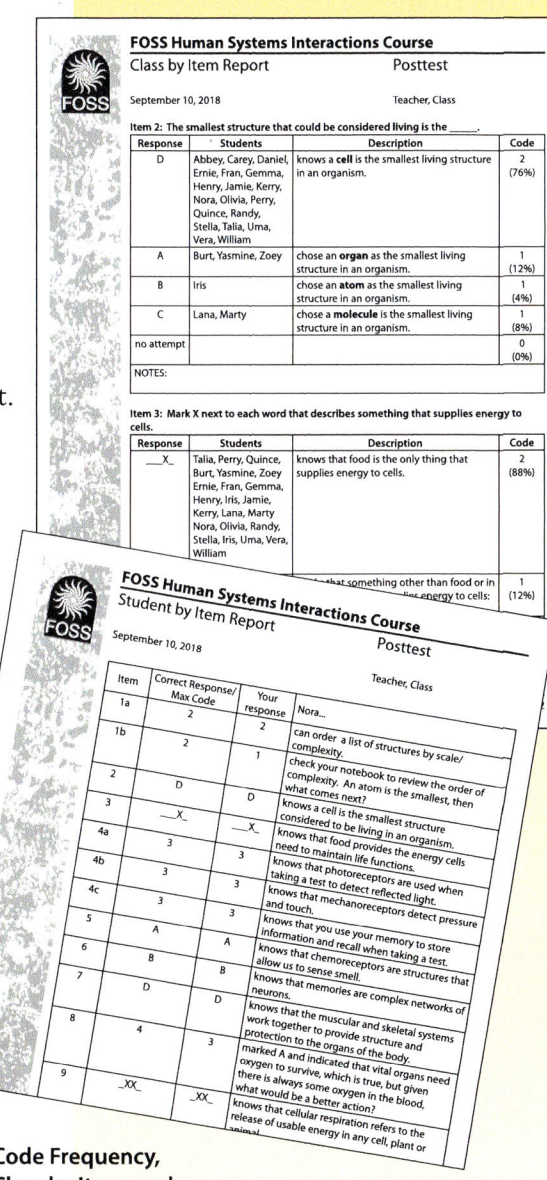

Code Frequency,
Class by Item, and
Student by Item Reports

Human Systems Interactions Course—FOSS Next Generation   **259**

HUMAN SYSTEMS INTERACTIONS — Assessment

ENTRY-LEVEL SURVEY

ENTRY-LEVEL SURVEY
HUMAN SYSTEMS INTERACTIONS

Name _____
Date _____ Class _____

NOTE: If you want to draw a picture to help you answer any of the items on this survey, ask your teacher for a piece of blank paper. Be sure to put your name on that extra piece of paper as well as this one.

1. What is the smallest living system of any organism? Explain why you think so.

   _____
   _____
   _____
   _____

2. In the table below, write the names of as many human body (organ) systems as you can in the column on the left. In the right column, briefly describe what that system does for the body.

Name of system	Function in the human body

3. How do cells get the energy they need to carry out life functions?

   _____
   _____
   _____
   _____
   _____

FOSS Next Generation
© The Regents of the University of California
Can be duplicated for classroom or workshop use.

Human Systems Interactions Course
Entry-Level Survey

Entry-Level Survey

Item 1

What to look for

Students might not think of a cell as a system. If they participated in the *FOSS Diversity of Life* course, they should know that cells are the basic units of life—the basic building blocks of all life. This item pushes students to explain what makes something living. For example, while atoms are smaller units, they are not a living system.

How this item can inform instruction

As you begin the "Structural Level Cards" activity in Investigation 1, Part 1, consider where students are in terms of understanding what constitutes a living system. For example, if students suggest that an atom (too small) or an organ (too complex) is the smallest living system, you'll want to clarify why a cell is the simplest system than can be considered a living thing (aka, single-celled organism).

Contributes to MS-LS1-1

Item 2

What to look for

Students might name individual organs and their functions rather than systems, such as the digestive or respiratory systems. This item will push students to try and identify the main body (organ) systems: digestive, circulatory (cardiovascular and lymphatic), respiratory, nervous, skeletal, muscular, excretory, and endocrine.

How this item can inform instruction

Consider student's knowledge of human systems and their functions. Use this information to determine how much you need to scaffold the discussion during the whole-class systems-interactions research in Investigation 1, Part 2. (We will not discuss the integumentary or reproductive systems in this course.)

Contributes to MS-LS1-3

Item 3

What to look for

Students most likely have misconceptions that cells get energy from things other than food, such as sleep, exercise, water, and caffeine. Students might not consider how organ systems are involved in getting food (or other perceived energy sources) to each cell in the body. In addition, students probably don't yet know about the role aerobic cellular respiration plays in enabling cells to release energy stored in food.

How this item can inform instruction

As you begin Investigation 2, use the insight you gain from this item to consider how you can help students move to the idea that only food can provide energy, how you might use models to explain how the digestive and circulatory systems work together to get food to each cell, and how cells release usable energy from food through the process of aerobic cellular respiration.

Contributes to MS-LS1-7

Human Systems Interactions Course—FOSS Next Generation

ENTRY-LEVEL SURVEY
HUMAN SYSTEMS INTERACTIONS

Name _____

4. Two students were talking. Student A was examining the fruit salad he was eating for lunch. He said, "Did you know that I am eating energy from the Sun?" Student B was eating a hamburger. She said, "I am eating energy that came from the Sun, too."

    Do you agree with these two students? Explain why you agree or disagree with each student.

    _____
    _____
    _____
    _____
    _____

5. Explain the stimulus-response process if you were to feel an ant crawling on your arm and you use your hand to brush it away.

    _____
    _____
    _____
    _____
    _____

6. How are memories formed?

    _____
    _____
    _____
    _____
    _____

Entry-Level Survey

Item 4

What to look for

Students probably have some experience with food webs and understanding that the Sun provides the energy captured by plants (producers) when they make food during the process of photosynthesis. That energy is passed on to consumers when they eat producers or other consumers. This item will provide evidence about how students think about energy and how it is passed from one organism to another.

How this item can inform instruction

Photosynthesis and the production of food is foundational to developing students' ideas about aerobic cellular respiration. They also sometimes confuse the two processes. Although there is a much stronger emphasis on aerobic cellular respiration in this course than photosynthesis, it is important to ensure that students understand the difference between the processes as well as how they are related: photosynthesis captures the energy; aerobic cellular respiration releases it.

Contributes to MS-LS1-7

Item 5

What to look for

Students might have some prior experience with how sensory receptors initiate a response to stimuli by sending messages to the brain. Most likely their answers will be incomplete, lacking details of how a message travels to and from the brain and/or the role of sensory receptors and different types of neurons.

How this item can inform instruction

As you begin Investigation 3, use the insight you gain from this item to consider what students understand about the nervous system and how you can build on those ideas as you explore how sensory receptors and neural pathways are involved with recognizing and responding to stimuli.

Contributes to MS-LS1-8

Item 6

What to look for

Students are not likely to have learned about the complex systems of memory formation and learning yet. Their answers may include some notion that the brain is involved in forming and storing memory, but might lack how neural networks are involved.

How this item can inform instruction

Understanding what students think about how memories are formed and stored will inform where you begin the exploration of how humans learn and form memories in Investigation 3, Part 4.

Contributes to MS-LS1-8

HUMAN SYSTEMS INTERACTIONS — *Assessment*

INVESTIGATIONS 1–2 I-CHECK

INVESTIGATIONS 1–2 I-CHECK
HUMAN SYSTEMS INTERACTIONS

ANSWERS

1. Write the letter of the organ system next to a matching function. Use each letter only once.

   - **E** gets rid of liquid waste from the body
   - **C** transports oxygen to the cells in the body
   - **H/A** gives the body structure and protection
   - **F** receives information from the environment
   - **H/A** enables the body to move
   - **G** uses hormones to communicate with the body
   - **B** turns food into nutrient molecules
   - **D** exhales waste gases such as carbon dioxide

   A skeletal system
   B digestive system
   C circulatory system
   D respiratory system
   E excretory system
   F nervous system
   G endocrine system
   H muscular system

2. Describe how organ systems interact to get *oxygen* to the cells.

   Cells get oxygen from the respiratory system. Oxygen enters our lungs and goes to the alveoli, which are surrounded by capillaries. Oxygen enters the bloodstream in the capillaries. The circulatory system carries oxygen to the cells.

3. Describe how organ systems interact to get *food* to the cells.

   When humans eat food, it goes into the digestive system and is broken down into food molecules. The food molecules enter the bloodstream from the small intestine and the circulatory system carries them to the cells.

Investigations 1–2 I-Check

Item 1

Focus on Disciplinary Core Ideas
Focus on Crosscutting Concepts

This item provides evidence that students can identify the functions of the major systems in the human body.

Code	If the student . . .
3	writes from top to bottom: E, C, H, F, A, G, B, D or E, C, A, F, H, G, B, D. (H and A can be interchanged in this item.)
2	marks all but one or two answers correctly.
1	marks any other answer.
0	makes no attempt.

▶ **ITEM 1 Next Steps**

Have students look through their notebooks for information about different body systems and discuss as a group. Compare group answers with the whole class. Discuss any lingering discrepancies. Have students revise their original responses as needed.

Item 2

Focus on Science and Engineering Practices
Focus on Disciplinary Core Ideas
Focus on Crosscutting Concepts

This item provides evidence that students have gathered and synthesized information that the body is a system of interacting subsystems; in this case, the delivery of oxygen to every cell.

Code	If the student . . .
3	refers to both the respiratory and circulatory systems *or* specific organs in those systems: e.g., states that blood (circulatory system or blood vessels such as capillaries) carries oxygen from the lungs (respiratory system) to cells.
2	includes minor errors such as one incorrect organ or refers to only one system.
1	writes any other answer.
0	makes no attempt.

▶ **ITEM 2 Next Steps**

Review the video *Circulatory and Respiratory System* available on FOSSweb. Discuss the item in small groups after viewing the video, and have students revise their answers as needed.

Item 3

Focus on Science and Engineering Practices
Focus on Disciplinary Core Ideas
Focus on Crosscutting Concepts

This item provides evidence that students have gathered and synthesized information that the body is a system of interacting subsystems; in this case, the delivery of food molecules to cells.

Code	If the student . . .
3	refers to both the digestive and circulatory systems *or* specific organs in those systems; e.g., states that blood (circulatory system or blood vessels such as capillaries) carries food or nutrient molecules from the small intestine (digestive system) to cells.
2	includes minor errors such as one incorrect organ or refers to only one system.
1	writes any other answer.
0	makes no attempt.

▶ **ITEM 3 Next Steps**

Have students review the video *Digestive and Excretory System* available on FOSSweb. Use the key-points strategy to develop a list of important information that students should have included in their answers. Have students review their own answers and revise as needed.

Human Systems Interactions Course—FOSS Next Generation

HUMAN SYSTEMS INTERACTIONS — *Assessment*

INVESTIGATIONS 1–2 I-CHECK
HUMAN SYSTEMS INTERACTIONS

ANSWERS

4. If blood cannot reach heart muscle cells, they can become damaged and die. What might happen to a person if some of his heart muscle cells die?

   Number the answers below to show the order in which structures would be affected if heart muscle cells die. If any phrase is not needed, leave it blank.

   __3__ Heart dies

   __4__ Circulatory system is unable to function and human dies

   _____ Heart atoms die

   __1__ Heart cells die

   __2__ Heart tissue dies

5. Explain how the circulatory system supports aerobic cellular respiration.

   The circulatory system carries oxygen and glucose to the cells, where they are used to produce usable energy for the cell through aerobic cellular respiration. The blood stream carries the waste products of cellular respiration (carbon dioxide and water) from the cells to the excretory system, where they leave the body.

6. The one thing that supplies energy to cells is ___*food (or glucose)*___.

FOSS Next Generation
© The Regents of the University of California
Can be duplicated for classroom or workshop use.

Investigations 1–2 I-Check

Item 4

Focus on Disciplinary Core Ideas
Focus on Crosscutting Concepts

This item provides evidence that the body is a system of interacting subsystems composed of structures of different complexities (scale) that can be affected by the health of other structures.

Code	If the student . . .
2	numbers phrases from top to bottom: 3, 4, _, 1, 2.
1	marks any other answer.
0	makes no attempt.

▶ **ITEM 4 Next Steps**

Have students consider their notes from the "Structural-Level-Cards" activity and discuss this item in small groups. Students then reflect and revise their answers as needed.

Item 5

Focus on Science and Engineering Practices
Focus on Disciplinary Core Ideas
Focus on Crosscutting Concepts

This item provides evidence that students understand the model explaining how the energy in matter (food) becomes usable energy for cells and how waste products are removed.

Code	If the student . . .
3	writes that the blood (circulatory system) carries oxygen and glucose *to* cells to produce energy, and carries carbon dioxide and water (waste products) *away* from the cells for disposal.
2	writes either that the blood carries oxygen and glucose to cells, or that the blood carries away carbon dioxide and water but not both, or includes both directions but does not mention specific resources, waste products, or energy; may state that blood disposes of waste products, rather than carries them away.
1	writes any other answer.
0	makes no attempt.

▶ **ITEM 5 Next Steps**

Have students refer to the model of aerobic cellular repiration in their notebooks. Use the key-points and revision-with-color strategies to help students reflect on this item. See the Next-Step Strategies section in this chapter for more information.

Item 6

Focus on Disciplinary Core Ideas
Focus on Crosscutting Concepts

This item provides evidence that students know only food supplies energy to cells.

Code	If the student . . .
2	writes "food" or "glucose."
1	writes anything else.
0	makes no attempt.

▶ **ITEM 6 Next Steps**

Have students check their science notebooks for information to confirm or refute their answers to this item. If that information is not there, take this opportunity to discuss with students what kinds of information are important to record.

Human Systems Interactions Course—FOSS Next Generation

INVESTIGATIONS 1–2 I-CHECK
HUMAN SYSTEMS INTERACTIONS

ANSWERS

7. A patient goes to the doctor with the following symptoms: trouble breathing, fatigue, hard time thinking clearly.

   a. Which organ system or systems are most likely affected by the patient's condition?
      (Mark the one best answer.)

      ○ A Nervous system only

      ○ B Respiratory system only

      ○ C Nervous and respiratory systems

      ● D All systems might be affected.

   b. Explain why you chose that answer.

      All the organ systems in the body are connected, so if one or two systems are affected by an illness, other systems may be affected as well. For instance, without oxygen, because the respiratory system is malfunctioning, cells throughout the body will die.

8. Aerobic cellular respiration occurs in cell structures called mitochondria. Cells with diseased mitochondria die. Why do those cells die?

   (Mark the one best answer.)

   ○ A Aerobic cellular respiration makes food for cells. Without food, the cells die.

   ○ B Aerobic cellular respiration makes carbon dioxide and food molecules. The cells becomes full of the extra molecules and die.

   ● C Aerobic cellular respiration provides usable energy for the cells. Without energy, the cells die.

   ○ D Aerobic cellular respiration provides cells with oxygen. Without oxygen, the cells die.

Investigations 1–2 I-Check

Item 7a
Focus on Disciplinary Core Ideas
Focus on Crosscutting Concepts

This item provides evidence that students know that the body is a system of interacting subsystems.

Code	If the student . . .
2	marks D.
1	marks any other answer.
0	makes no attempt.

▶ **ITEM 7ab Next Steps**

Use the review-and-critique anonymous-student-work strategy if students need help reflecting on this item. See the Next-Step Strategies section in this chapter for more information.

Item 7b
Focus on Science and Engineering Practices
Focus on Disciplinary Core Ideas
Focus on Crosscutting Concepts

This item provides evidence that students can argue with evidence that the body is a system of interacting subsystems that can affect each other.

Code	If the student . . .
3	explains that all organ systems in the body are connected, and includes an example.
2	refers to only one or two systems besides the respiratory and nervous systems as being affected by the symptoms.
1	writes any other answer.
0	makes no attempt.

Item 8
Focus on Science and Engineering Practices
Focus on Disciplinary Core Ideas
Focus on Crosscutting Concepts

This item provides evidence that students are able to gather and synthesize information to explain that certain diseased cellular structures can cause disruptions in the flow of energy through an organism.

Code	If the student . . .
2	marks C.
1	marks any other answer.
0	makes no attempt.

▶ **ITEM 8 Next Steps**

If the class is equally divided in their answers, or only a few students have the correct answer, use the multiple-choice-corners strategy to reflect on this item. Or you can use the multiple-choice-discussion strategy. See the Next-Step Strategies section in this chapter for more information.

Human Systems Interactions Course—FOSS Next Generation

INVESTIGATIONS 1–2 I-CHECK
HUMAN SYSTEMS INTERACTIONS

ANSWERS

9. Mark **X** next to each sentence that helps describe how the *respiratory* system and the *circulatory* system interact in humans.

   _____ The two systems don't interact.

   __X__ Blood picks up oxygen in the lungs.

   _____ Blood picks up nutrients from the small intestine.

   __X__ Blood carries carbon dioxide to the lungs.

   __X__ Blood carries nutrients to cells in the organs of the respiratory system.

   _____ Blood cells are manufactured in the lungs.

10. Aplastic anemia is a disease that affects the production of red blood cells in the bone marrow of leg and arm bones. One of the symptoms of aplastic anemia is severe headaches. Explain why that might be so.

    Red blood cells carry resources such as oxygen to the cells in the body, including cells in the brain. If the production of red blood cells decreases, the brain may not receive enough oxygen or other resources, potentially causing the headaches.

Investigations 1–2 I-Check

Item 9

Focus on Science and Engineering Practices
Focus on Disciplinary Core Ideas
Focus on Crosscutting Concepts

This item provides evidence that students have developed a model that explains how systems in the human body interact; in this case, the interaction of the respiratory and circulatory systems.

Code	If the student...
3	marks the second, fourth, and fifth statements.
2	marks as for score 3, but also includes the third statement (which is a true statement, but is not part of the respiratory/circulatory interaction).
1	marks any other way.
0	makes no attempt.

▶ **ITEM 9 Next Steps**

Review the readings "Respiratory System" and "Circulatory System" in *FOSS Science Resources* and discuss needed revisions in small groups.

Item 10

Focus on Science and Engineering Practices
Focus on Disciplinary Core Ideas
Focus on Crosscutting Concepts

This item provides evidence that students can synthesize information and apply the model they have developed that the body is a system of interacting subsystems to explain how problems in one system can result in symptoms in other systems.

Code	If the student...
3	writes that red blood cells carry oxygen (or resources in general) and that if there are not enough red blood cells, the brain cells will not receive enough oxygen, leading to severe headaches.
2	writes that headaches might be caused by decreased blood flow; does not make the connection to decreased oxygen or resources.
1	writes any other answer.
0	makes no attempt.

▶ **ITEM 10 Next Steps**

Develop a list of the key points to help students reflect on this item, then use the review-and-critique-anonymous-student-work strategy before students revise their answers. See the Next-Step Strategies section in this chapter for more information.

Human Systems Interactions Course—FOSS Next Generation

HUMAN SYSTEMS INTERACTIONS — Assessment

INVESTIGATION 3 I-CHECK

INVESTIGATION 3 I-CHECK
HUMAN SYSTEMS INTERACTIONS

ANSWERS

1. Write the first letter of the sense next to each phrase it matches. You may use each sense more than once.

   __S__ uses chemoreceptors
   __T__ allows a person to sense pain
   __V__ uses rods and cones
   __S__ responds to molecules in the air
   __V__ uses photoreceptors
   __T__ uses mechanoreceptors
   __V__ responds to light

   Key
   T Touch
   S Smell
   V Vision (sight)

2. Spinal injuries can cause loss of sensation and paralysis (loss of ability to move) in the legs. Explain why this can happen.

   (Inter)neurons in the spinal cord carry messages between the brain and the legs. If the spinal cord is damaged, sensory messages will not be able to go to the brain and the "movement" messages may not get to the muscles in the legs. So the person won't be able to feel or move her legs.

3. Your lips are more sensitive to touch than your forehead. What is different about the touch receptive fields in the lips compared to the forehead?
   (Mark the one best answer.)

   ● A There are more touch receptive fields in the lips.
   ○ B The touch receptive fields are bigger in the lips.
   ○ C The touch receptive fields are more sensitive in the lips.
   ○ D The touch receptive fields are used more often in the lips.

   FOSS Human Systems Interactions Course, Second Edition
   © The Regents of the University of California
   Can be duplicated for classroom or workshop use.

   Human Systems Interactions Course
   Investigation 3 I-Check
   Page 1 of 4

Investigation 3 I-Check

Item 1

Focus on Disciplinary Core Ideas
Focus on Crosscutting Concepts

This item provides evidence that students can identify which sensory receptors (structures) respond to particular stimuli (functions).

Code	If the student . . .
2	writes from top to bottom: S, T, V, S, V, T, V.
1	writes anything else.
0	makes no attempt.

▶ **ITEM 1 Next Steps**

Use the group consensus/whiteboard strategy to help students reflect on this item. See the Next-Step Strategies section in this chapter for more information.

Item 2

Focus on Science and Engineering Practices
Focus on Disciplinary Core Ideas
Focus on Crosscutting Concepts

This item provides evidence that students can synthesize and apply information about how the nervous system works to explain the effects of a spinal injury.

Code	If the student . . .
3	writes that damage to the spinal cord will interrupt messages carried between the brain and the legs; includes both that a person won't be able to sense anything because messages can't get to the brain and a person won't be able to move her legs because a message can't get from the brain to the legs; may include interneurons.
2	writes that damage to the spine will interrupt messages either from the legs or from the brain, but not both.
1	writes any other answer.
0	makes no attempt.

▶ **ITEM 2 Next Steps**

In small groups, have students discuss their answers. They can also review "Brain Messages" in *FOSS Science Resources*. Reconvene the class and have, each group report about something they agreed upon.

Item 3

Focus on Disciplinary Core Ideas
Focus on Crosscutting Concepts

This item provides evidence that students understand the structure and function of receptive fields that cause greater sensitivy of touch in lips compared to the forehead.

Code	If the student . . .
2	marks A.
1	marks any other answer.
0	makes no attempt.

▶ **ITEM 3 Next Steps**

Point students to notebook sheet 7, *Touch*, to review information about touch receptors. You can also project teacher master C, *Pressure Receptive-Fields Models A*, for reference. Students review the information, reflect, and revise their answers as needed.

Human Systems Interactions Course—FOSS Next Generation

INVESTIGATION 3 I-CHECK
HUMAN SYSTEMS INTERACTIONS

ANSWERS

4. Study the image below.

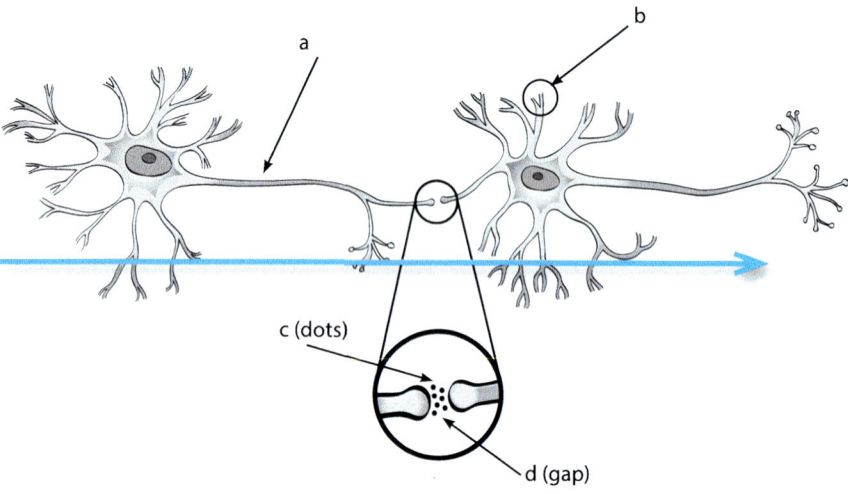

Write the name of each structure on the lines below indicated by a letter in the image above.

a. _____axon_____

b. _____dendrite_____

c. _neurotransmitters_

d. _____synapse_____

e. Draw an arrow on the image above that indicates the direction a neural message is transmitted. _See arrow on diagram above._

5. Neurotransmission has both electric and chemical components. Explain why both are needed.
The axon carries an electric message along a neuron, to the end of the axon. At the synapse, the electric message stimulates chemicals called neurotransmitters to cross the gap, and the electric message continues along the next neuron.

Investigation 3 I-Check

Item 4a–d
Focus on Disciplinary Core Ideas
Focus on Crosscutting Concepts

This item provides evidence that students can identify the structures of neurons.

Code	If the student . . .
2	correctly labels all four structures in the image.
1	writes any other answer.
0	makes no attempt.

▶ **ITEM 4a-e Next Steps**
Students can review the "Brain: Synapse Function" activity available on FOSSweb and compare it to their diagram of two neurons in their science notebooks. Then have students reflect on and revise their labels on the diagram.

Item 4e
Focus on Disciplinary Core Ideas
Focus on Crosscutting Concepts

This item provides evidence that students know the direction in which a message is sent across neurons (function).

Code	If the student . . .
2	draws an arrow from left to right (from axon to dendrite).
1	draws any other arrow.
0	makes no attempt.

Item 5
Focus on Science and Engineering Practices
Focus on Disciplinary Core Ideas
Focus on Crosscutting Concepts

This item provides evidence that students can apply information about how neurons transmit signals to explain the functions of electric and chemical components of neurotransmission.

Code	If the student . . .
3	includes correct descriptions of both an electric message traveling along a neuron and chemicals (neurotransmitters) crossing the gap (synapse) between neurons.
2	includes either correct description of an electric component within a neuron or a chemical component between neurons.
1	implies that the message that crosses the synapse is electric, not chemical, or writes any other answer.
0	makes no attempt.

▶ **ITEM 5 Next Steps**
Have students review the "Brain: Synapse Function" activity available on FOSSweb and their reflections from creating a neurotransmitter model. Use the revision-with-color strategy to improve answers. See the Next-Step Strategies section in this chapter for more information.

Human Systems Interactions Course—FOSS Next Generation 275

INVESTIGATION 3 I-CHECK
HUMAN SYSTEMS INTERACTIONS

ANSWERS

6. A person who is paralyzed can use an electronic device implanted in the brain to control robotic hands. How is that possible?

   Neural messages are electric impulses that can stimulate an electronic device if it is connected to neurons in the brain. The brain can send a signal through the device to move a robotic hand.

7. You have an experience that causes a complex network of neurons to be built in your brain. This network is most likely a _____.
   (Mark the one best answer.)

   ○ **A** receptive field
   ○ **B** disease
   ● **C** memory
   ○ **D** motor pathway

8. Neurotransmitters _____.
   (Mark the one best answer.)

   ○ **A** carry electric impulses along a neuron
   ○ **B** carry chemical impulses along a neuron
   ● **C** are chemicals that move across synapses
   ○ **D** are chemicals that carry electric impulses across synapses

Investigation 3 I-Check

Item 6

Focus on Science and Engineering Practices
Focus on Disciplinary Core Ideas
Focus on Crosscutting Concepts

This item provides evidence that students can apply knowledge about the nervous system to explain how an electronic device implanted in the brain can control (cause and effect) a robotic hand (structure/function).

Code	If the student . . .
3	states that neural messages are electric impulses or signals; explains that the electronic device must receive electric impulses from neurons in the brain and then the implanted device sends a signal to move the robotic hand.
2	mentions anything about electric messages.
1	mentions neurotransmitters with no other information, or writes anything else.
0	makes no attempt.

▶ **ITEM 6 Next Steps**

Use the class debate next-step strategy if students need help reflecting on this item. See the Next-Step Strategies section in this chapter for more information.

Item 7

Focus on Disciplinary Core Ideas
Focus on Crosscutting Concepts

This item provides evidence that students understand that experiences can cause a complex network of neurons (structures) in the brain to create a memory (function).

Code	If the student . . .
2	marks C.
1	marks any other answer.
0	makes no attempt.

▶ **ITEM 7 Next Steps**

If students need help with this item, use the multiple-choice corners strategy. See the Next-Step Strategies section in this chapter for more information.

Item 8

Focus on Disciplinary Core Ideas
Focus on Crosscutting Concepts

This item provides evidence that students know that neurotransmitters are chemicals that move impulses across synapses, but do not move along a neuron or carry electric impulses along a neuron (structure/function).

Code	If the student . . .
2	marks C.
1	marks any other answer.
0	makes no attempt.

▶ **ITEM 8 Next Steps**

Use the multiple-choice discussion strategy to help students reflect on and revise their answers. See the Next-Step Strategies section in this chapter for more information.

Human Systems Interactions Course—FOSS Next Generation

HUMAN SYSTEMS INTERACTIONS — Assessment

INVESTIGATION 3 I-CHECK
HUMAN SYSTEMS INTERACTIONS

ANSWERS

9. Your hand touches something super gooey and icky. In the space below, draw a flowchart or diagram to describe the steps in the neural pathway that leads to pulling your hand out of the goo. Start with the stimulus. Write and draw your answer.

Written answers will vary, but should be a description of the diagram shown below.

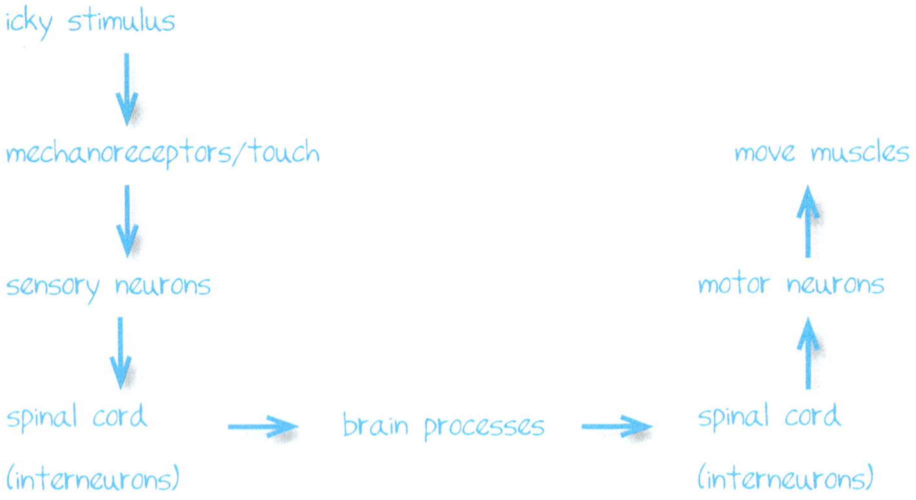

icky stimulus
↓
mechanoreceptors/touch
↓
sensory neurons
↓
spinal cord (interneurons) → brain processes → spinal cord (interneurons)
↑
motor neurons
↑
move muscles

Investigation 3 I-Check

Focus on Science and Engineering Practices
Focus on Disciplinary Core Ideas
Focus on Crosscutting Concepts

Item 9

This item provides evidence that students have developed a reasonable model that explains the steps in the neural pathway from stimuli to reaction (structure and function; cause and effect).

Code	If the student . . .
4	correctly reports all the steps in the pathway from stimulus to movement (see sample answer).
3	misses a step or does not include all the vocabulary; may not include the stimulus.
2	includes major misconceptions (e.g., missing the brain-processing information or moving the message in the incorrect direction).
1	writes any other answer.
0	makes no attempt.

▶ **ITEM 9 Next Steps**
Review notebook sheets 8–9, *Neural Pathways and Response Sheet—Investigation 3*. Use the group-consensus/whiteboards strategy to share group ideas as a class. See the Next-Step Strategies section in this chapter for more information.

Human Systems Interactions Course—FOSS Next Generation

POSTTEST
HUMAN SYSTEMS INTERACTIONS

ANSWERS

1. a. Number the following in order from least complex (1) to most complex (4) structure.

    __1__ Muscle cell

    __3__ Heart

    __2__ Heart tissue

    __4__ Circulatory system

   b. Number the following in order from least complex (1) to most complex (4) structure.

    __2__ Molecule

    __4__ Cell

    __1__ Atom

    __3__ Cell structure (organelle)

2. The smallest structure that could be considered living is the _____.
   (Mark the one best answer.)

   ○ A organ

   ○ B atom

   ○ C molecule

   ● D cell

3. Mark **X** next to each word that describes something that supplies energy to cells.

   _____ Water

   _____ Exercise

   _____ Sleep

   __X__ Food

   _____ Caffeine

FOSS Next Generation
© The Regents of the University of California
Can be duplicated for classroom or workshop use.

Human Systems Interactions Course
Posttest
Page 1 of 4

Posttest

Item 1a
Focus on Disciplinary Core Ideas
Focus on Crosscutting Concepts

This item provides evidence that students understand the complexity and scale of structures that make up human body systems.

Code	If the student...
2	writes from top to bottom: 1, 3, 2, 4.
1	numbers any other way.
0	makes no attempt.

Item 1b
Focus on Disciplinary Core Ideas
Focus on Crosscutting Concepts

This item provides evidence that students understand the complexity and scale of structures that make up human body systems.

Code	If the student...
2	writes from top to bottom: 2, 4, 1, 3.
1	numbers any other way.
0	makes no attempt.

Item 2
Focus on Disciplinary Core Ideas
Focus on Crosscutting Concepts

This item provides evidence that students know a cell (structure) is the smallest unit that can be said to be alive (function).

Code	If the student...
2	marks D.
1	marks anything else.
0	makes no attempt.

Item 3
Focus on Disciplinary Core Ideas
Focus on Crosscutting Concepts

This item provides evidence that students know only food supplies energy to cells.

Code	If the student...
2	marks food only.
1	marks any other way.
0	makes no attempt.

Human Systems Interactions Course—FOSS Next Generation

POSTTEST   ANSWERS
HUMAN SYSTEMS INTERACTIONS

4. You are taking a test at this moment. Describe how the following are involved.

   a. Photoreceptors

   Light travels from the paper to my eyes. Light stimulates photoreceptors.

   b. Mechanoreceptors

   My fingers feel the pencil, and my hand is resting on the paper. The sense of touch uses mechanoreceptors.

   c. Memory

   I can remember and respond to a specific question because I built the memory when I studied.

5. Scientists have found that chemoreceptors have strong connections to the hippocampus (the part of the brain involved in emotion and memory).

   This means that strong memories or emotions can be stimulated by _____.
   (Mark the one best answer.)

   ○ A   touch
   ● B   smells
   ○ C   images
   ○ D   thoughts

Posttest

Focus on Science and Engineering Practices
Focus on Disciplinary Core Ideas
Item 4abc
Focus on Crosscutting Concepts

This item provides evidence that students can apply information synthesized from informational text that different sensory receptors send different messages to the brain for immediate behavior or storage as memories (cause and effect; structure and function).

Code	If the student . . .
Item 4a	Photoreceptors
3	writes that light from the test is entering the eyes and hitting the photoreceptors, allowing the student to see.
2	writes that the student can see or read; does not mention light as the stimulus.
1	writes anything else.
0	makes no attempt.
Item 4b	Mechanoreceptors
3	writes that the student can feel the pencil and/or paper due to mechanoreceptors (pressure or touch receptors) in the skin; might mention hearing.
2	writes that mechanoreceptors in the skin help the student feel and hold the pencil.
1	writes anything else.
0	makes no attempt.
Item 4c	Memory
3	refers to having formed a memory (when studying or learning), which can now be recalled.
2	writes a general statement about memory recall without reference to memory formation.
1	writes anything else.
0	makes no attempt.

> **NOTE**
> Score Item 4abc as three separate items. The coding guide is condensed here to save space.

Focus on Disciplinary Core Ideas
Item 5
Focus on Crosscutting Concepts

This item provides evidence that students can apply knowledge that chemoreceptors allow us to sense smells (structure/function).

Code	If the student . . .
2	marks B.
1	marks any other answer.
0	makes no attempt.

Human Systems Interactions Course—FOSS Next Generation

POSTTEST
HUMAN SYSTEMS INTERACTIONS

ANSWERS

6. Memories are _____.
   (Mark the one best answer.)
   - ● **A** complex networks of neurons
   - ○ **B** weakened connections between synapses in the brain
   - ○ **C** two neurons connected together
   - ○ **D** complex electric signals that communicate

7. The muscular and skeletal systems work together to _____.
   (Mark the one best answer.)
   - ○ **A** dispose of liquid waste from the body
   - ○ **B** pump oxygen into the bloodstream
   - ○ **C** transmit information from the environment to the nervous system
   - ● **D** provide structure and protection to the organs of the body

8. • You witness an accident. The injured person is not breathing and has no pulse. The most important action you can take is to call 911, and then _____.
   (Mark the one best answer.)
   - ○ **A** give the patient rescue breaths (breathing air into the lungs)
   - ○ **B** elevate their legs to send the blood back to the heart
   - ● **C** use chest compressions to keep the heart pumping blood
   - ○ **D** try to shake the patient awake and keep him or her talking

   • Why would this be the most important action to take?
   There is always dissolved oxygen in the blood, so the most important thing is to keep the blood flowing. Blood brings oxygen to the brain and other vital organs to keep them alive.

Posttest

Item 6
Focus on Disciplinary Core Ideas
Focus on Crosscutting Concepts

This item provides evidence that students know memories are complex networks of neurons that can change over time (systems and system models; structure/function).

Code	If the student ...
2	marks A.
1	marks any other answer.
0	makes no attempt.

Item 7
Focus on Disciplinary Core Ideas
Focus on Crosscutting Concepts

This item provides evidence that students can identify the functions and interactions of the skeletal and muscular systems.

Code	If the student ...
2	marks D.
1	marks any other answer.
0	makes no attempt.

Item 8
Focus on Science and Engineering Practices
Focus on Disciplinary Core Ideas
Focus on Crosscutting Concepts

This item provides evidence that students can apply what they know about the structure and function of body systems (especially the circulatory and respiratory), in order to evaluate the best solution to help an injured person.

Code	If the student ...
4	marks C and writes that blood carries oxygen to organs to keep them alive; chest compressions keep the blood circulating.
3	marks A and writes that the vital organs need oxygen to survive, so rescue breaths add oxygen to the body.
2	marks C or A, but explanation is vague or unconvincing.
1	writes anything else.
0	makes no attempt.

Human Systems Interactions Course—FOSS Next Generation

POSTTEST
HUMAN SYSTEMS INTERACTIONS

ANSWERS

9. When scientists refer to *cellular respiration*, they include which of the following processes?
   (You may mark as many of the statements below as needed to answer the question.)

   _____ When air is breathed in and circulated throughout the body

   __X__ When plant cells use glucose to release usable energy for the cells

   __X__ When usable energy is released in a cell to carry out necessary functions

   _____ When plants transfer water from the ground to all the plant's cells

10. Student A and Student B were having a debate.

    Student A says, "A disease that affects the respiratory system can also affect the nervous system."

    Student B says, "That's not right. If you have allergies, it's harder to breath, but nothing else is affected."

    If you were part of this debate, whose side would you take? Be sure to include evidence that supports the side you want to argue for.

    Student takes Student A's side.
    The respiratory system is important because it brings oxygen into the body, but the circulatory system delivers the oxygen to the cells in the body, including the brain and nerve cells. If the brain and nerve cells (part of the nervous system) are not getting enough oxygen, then the nervous system may start to fail.

Posttest

Item 9
Focus on Disciplinary Core Ideas
Focus on Crosscutting Concepts

This item provides evidence that students know that the function of aerobic cellular respiration is to provide energy for cells.

Code	If the student...
2	marks the second and third statements.
1	marks any other way.
0	makes no attempt.

Item 10
Focus on Science and Engineering Practices
Focus on Disciplinary Core Ideas
Focus on Crosscutting Concepts

This item provides evidence that students can argue with evidence that the body is a system of interacting subsystems.

Code	If the student...
3	takes Student A's side; includes that all the organ systems interact and in this particular case, the respiratory system provides oxygen for the cells in the entire body, so if any cells in the nervous system do not receive oxygen, they will be affected.
2	takes Student A's side, but argument is flawed or vague.
1	takes Student B's side.
0	makes no attempt.

Human Systems Interactions Course—FOSS Next Generation

HUMAN SYSTEMS INTERACTIONS — Assessment

BENCHMARK ASSESSMENT
Summary Chart

	Entry-Level Survey	I-Check 1–2	I-Check 3	Posttest
Structure and Function				
Concept A All living things need food, water, a way to dispose of waste, and an environment in which they can live.				
• The cell is the basic unit of life. All living things are one or more cells.	1			2
• Aerobic cellular respiration is the process by which energy stored in food molecules is converted into usable energy for cells.	3	5, 6, 8		3, 9
• In multicellular organisms, such as humans, cells form tissues, tissues form organs, and organs form organ systems (subsystems), which interact to serve the needs of the organism.		4		1ab
• The human body is a system of interacting subsystems (circulatory, digestive, endocrine, excretory, muscular, nervous, respiratory, skeletal).	2	1, 2, 3, 7ab, 9, 10		7, 8, 10
Concept C Animals detect, process, and use information about their environment to survive.				
• The nervous system is a human subsystem that functions to gather and synthesize information from the environment.			2, 4a–e, 6	
• Sensory receptors are structures that respond to stimuli by sending messages to the brain for processing and response.	5		1, 3, 5, 8, 9	4a–c
• Neural pathways change and grow as information is acquired and stored as memories.	6		7	5, 6
Complex Systems				
Concept A Organisms and populations of organisms depend on their environmental interactions with other living things and with nonliving factors.				
• The Sun provides energy that plants use to produce food molecules from carbon dioxide and water. The energy in food molecules is processed in the cells of most organisms to drive life processes.	4			
Science and Engineering Practices	1, 3, 6	2, 3, 5, 7b, 8, 9, 10	2, 5, 6, 9	4a–c, 8, 10
Crosscutting Concepts	1, 2, 3, 4, 5, 6	1, 2, 3, 4, 5, 6, 7ab, 8, 9, 10	1, 2, 3, 4a–e, 5, 6, 7, 8, 9	1ab, 2, 3, 4a–c, 5, 6, 7, 8, 9, 10

TEACHER NOTES

TEACHER NOTES